Latrinae

Roman Toilets in the Northwestern Provinces of the Roman Empire

edited by

Stefanie Hoss

Archaeopress Roman Archaeology 31

Archaeopress Publishing Ltd
Summertown Pavilion
18-24 Middle Way
Oxford OX2 7LG

www.archaeopress.com

ISBN 978 1 78491 725 8
ISBN 978 1 78491 726 5 (e-Pdf)

© Archaeopress and the authors 2018

The cover image was inspired by the wooden toilet seat excavated in Vindolanda in August 2014 (see www.vindolanda.com). It is believed to be the only surviving wooden seat in the Roman Empire, but it is also probably the only known toilet seat with a hole that is oval, not round. According to the excavator, Dr Andrew Birley, the Vindolanda seat had been used for quite some time and then was decommissioned and discarded amongst the rubbish left behind in the final fort at the site (before the construction of Hadrian's Wall started in the early second century), where it was preserved in the anaerobic, oxygen free, conditions.

All rights reserved. No part of this book may be reproduced, in any form or
by any means, electronic, mechanical, photocopying or otherwise,
without the prior written permission of the copyright owners.

This book is available direct from Archaeopress or from our website www.archaeopress.com

An artist's impression of the inside of the latrine at Bearsden with a soldier holding moss prior to cleaning himself. Drawing by Michael J. Moore

Contents

Introduction 1
Stefanie Hoss

Sewers or cesspits? Modern assumptions and Roman preferences 5
Gemma Jansen

The latrine at the Roman fort on the Antonine Wall at Bearsden 19
David J Breeze

Flushed with success – a Roman flushing installation in the latrines of the Great Bathhouse of the Colonia Ulpia Traiana near Xanten (D) 23
Norbert Zieling

The latrines of Roman Aachen 29
Andreas Schaub

An outhouse in the garden? – Looking at a backyard in the *vicus* of Bonn 35
Jeanne-Nora Andrikopoulou-Strack, Manuel Fiedler and Constanze Höpken

A bath with public toilets in the *vicus* of Bonn 43
Gary White

The Roman public toilet of Rottenburg am Neckar 47
Stefanie Hoss

Latrines connected to bathhouses in Germania inferior – an overview 55
Michael Dodt

Roman toilets in Nijmegen, *Oppidum Batavorum* and *Ulpia Noviomagus*, the Netherlands ... 77
Elly N.A. Heirbaut

Arlon, apport des découvertes récentes dans le vicus à l'examen des latrines gallo-romaines 89
Denis Henrotay

A Roman latrine near St. Kolumba in Cologne and its remarkable contents 95
Michael Dodt

Latrine pits in the Roman *vicus* of *Vitudurum* / Oberwintherthur (Switzerland) 103
Verena Jauch

A Roman cesspit from the mid-2nd century with lead price tags in the civil town of Carnuntum (Schloss Petronell/Austria) 119
Beatrix Petznek

Roman chamber pots 127
Beatrix Petznek

A Roman 'Toilet bowl' from Speicher (Eifelkreis Bitburg-Prüm, Rhineland-Palatinate, Germany) 137
Bernd Bienert

The meaning of *stercus* in Roman military papyri – dung or human faeces? Or: who is supposed to clean *this* shit up? 143
Kai Juntunen

Introduction

Stefanie Hoss

Buildings can be classed into two groups: one – very large – group that can be termed multi-purpose and that is used with little difference in design for the many and quite variable tasks of living and working and another, smaller group of buildings that are built for a specific purpose clearly defined through their form or installations. This of course does not exclude the use of these buildings for other functions, but is an indication of their main intended use. In the Roman period, examples of this group include many technical buildings, such as aqueducts, harbour installations and defensive walls, but also buildings used by humans every day, such as bath buildings and toilets. While the technical buildings were mainly functional, bath buildings were hubs of the social life of a settlement and toilets also must have seen social exchanges.

During the last four decades, the interest in the social activities that make up the experience of daily life in the Roman Empire has increased both in archaeological scholarship as well as the general public. This had also boosted the interest in Roman toilets.

Before the latter part of the 20th century, Roman toilets and their infrastructure had often been overlooked, and during the 19th century they were occasionally even misinterpreted. This was a result of both the inclination of early archaeologists to focus on the high art of the Roman period and leave daily life unexplored, and the taboos of the age, which inhibited writing about bodily functions. This changed in the 20th century and Roman toilets were recognized and described for what they were. Still, the descriptions of the installations often were very superficial ('a latrine with the usual installations,) and toilets were not analysed systematically.

When Gemma Jansen started her research on the water supply of Herculaneum in 1989, it quickly emerged that the many and well preserved toilets of this city and of its neighbour Pompeii had never been studied thoroughly and she realised that she had found a whole new research area.

One would expect that in these modern times, the taboo around Roman toilets would have vanished, but in a surreptitious manner it seems to still influence research. This is proven by the giggly reactions of many colleagues when we organised a conference on Roman Toilets in the Northwestern provinces of the Roman Empire and sent around a 'Call for Papers'. It also is present in the obvious distaste for the subject exhibited by some eminent archaeologists and the gleeful coverage of the conference in the press.

Fortunately, this attitude seems to change with a growing interest in the 'grittier aspects' (Draycott 2012; Thüry 2001) of daily life in the Roman world, which has sparked a wealth of new research into dirt, pollution and sanitation as well as waste management and disposal (Bradley 2012; Hobson 2009; Koloski-Ostrow 2015; Magness 2011; Magness 2012; Neudecker 1994; Raventós and Remolà 2000; Scobie 1986; Thüry, 2001).

A sign of these new times was the first conference on Roman toilets and sanitation, organized 2007 in Rome by Gemma C.M. Jansen, Ann Olga Koloski-Ostrow and Eric. M. Moorman (Jansen *et al.* 2011). The conference was very stimulating and satisfying in many aspects, but it was focused on the Roman toilets in the Mediterranean. Because my own research focus lies in the Northwestern provinces of the Roman Empire, I knew that many interesting examples of multi-seater channel toilets from that region were still either unpublished or relatively unknown among archaeologists. Elly Heirbaut, who had just led an excavation of a part of *oppidum Batavorom* in Nijmegen with lots of cesspit toilets was in a similar situation with regard to cesspit toilets, and we decided on the spot that we would organise our own conference - something that we can (now) heartily recommend for other researchers in a similar situation.

The planning of the conference took two years, some daunting grant applications and the encouragement and help of friendly mentors. Gemma Janssen and Eric Moorman were especially helpful and we would like to thank them here. We also greatly benefited from the practical help given to us for free by two (then) students of the Radboud University, Suzanne van der Liefvoort and Klaartje Huijben, who helped us during the preparations and took over many tasks throughout the conference. Both their cheerfulness and their problem-solving capacities seemed endless and it was a relief for us to be able to concentrate on the proceedings of the conference instead of the logistics.

We were lucky to gain the financial support of the municipality of Nijmegen (Gemeente Nijmegen), the Radboud University and the Gerda-Henkel-Stiftung Düsseldorf. Their generous financial aid and the logistical help of the Radboud University made it possible for us invite lecturers, which greatly helped in attracting interesting papers. And having the possibility to offer everybody attending free coffee seems to have supported an animated exchange of ideas during the breaks.

The conference was focused on the toilets of the Northwestern provinces of the Roman Empire and consisted of two days with different structures and different participants. It took place at the campus of the Radboud University on the 1st and 2nd of May 2009. The first day was structured into sets of lectures and directed at archaeologists, archaeo-botanists and archaeo-zoologists as well as the interested public.

The proceedings were opened by a keynote lecture by Gemma Jansen, in which she contrasted channel toilets and cesspit toilets in the Mediterranean (see contribution Jansen in this volume). There followed fourteen case studies, grouped into four sessions, starting with the public channel toilets and military toilets in the morning and continuing with cesspit toilets, chamber pots and various scientific research methods in the afternoon. Collectively, the papers could unanimously show that the inhabitants of the Northwestern provinces used the whole bandwidth of possibilities for the collection and removal of human waste which had earlier been available in the Mediterranean: large multi-seater channel toilets, cesspit toilets and portable toilet bowls or chamber pots. Most contributions stimulated lively debate, which often spilled over into the coffee breaks. The day concluded with a nice dinner in town.

On the second day, a smaller group of excavators reconvened at the campus of the Radboud University to discuss cesspit toilets and public multi-seater toilets with channels in two workshops. The workshops concentrated on the manifold problems of excavating Roman toilets and were organized as round table discussions – with the plans of several excavations on the table. The idea was to gather practical hints and approaches for the quick identification and correct methods of excavation and sampling of Roman toilets.

During the very animated discussions, a variety of issues were debated. While the previous day had demonstrated the value of scientific analysis of waste to understand the food habits and diseases of the Roman users of the toilet, on the second day, questions on how to find the necessary expertise and financing for such investigations were raised. Another point was the necessity in modern excavations – especially developer-funded excavations – to have a quick and cheap method by which to differentiate the enormous amount of pits often found between 'mere' rubbish-pits to be excavated quickly (as their main interest is their content) and cesspits, which should be excavated with more care, in order to be able to evaluate not only the content, but also the manner of construction of the cesspit and possibly its superstructure.

The method suggested by Denis Henrotay was based on taking samples from the earth on the inside rim of the pit and mixing it with a little water. This solution is then tested with ph soil test strips easily available in garden centres. If the soil has a phosphorus level that is higher than that of the ambient soil, the pit had likely been filled with human waste.

Both the first day of lectures and the second day of workshops could not have taken place or been as stimulating without the contributions of all the participants of the conference, whether they presented a case study, contributed to the discussion in the workshops or simply asked a question and we are deeply appreciative of that.

Turning the conference into a book has been another long drawn-out challenge. Unfortunately, not all of the original contributors were able to submit articles, but during the long process, other researchers were approached and asked if they would be interested in submitting an article. These articles now make up a third of the book.

We owe thanks to all the contributing authors for their patience and fortitude in waiting for the publication for some years and also for enduring the long editing process that was entailed by the translation of the articles.

The case studies on Roman toilets presented in this book come from Austria, Belgium, Germany, Great Britain, the Netherlands and Switzerland. While some subjects have been published elsewhere – with the exception of the case study of Bearsden – they are presented in English for the first time.

The first paper is the only one not dealing with toilets in the Northwestern provinces, but with our expectations of Roman toilets in the Mediterranean. Jansen explains that those toilet we often deem 'typical' Roman toilets – namely stone channel toilets with underground sewer systems – were only one possible solution for the problem of waste removal in the Mediterranean, and the most expensive one at that. Wherever possible, the much cheaper and simpler cesspits were used, proving that this is as much a 'Roman' type of toilet as the channel toilets.

The second part of the book is made up of case studies of stone channel toilets. The contexts of these are quite mixed, from a smallish installation at in the fort of Bearsden on the Antonine Wall in Britain (Breeze), or two equally small outhouses in Bonn / Germany

(Andrikopoulou *et al.* and White) to a large installation in a public bathhouse Xanten (Zieling) and an installation unconnected to a bathhouse in Rottenburg (Hoss). While these papers all present a single toilet, two papers are overviews of a number of stone channel toilets in a single city (Schaub) or connected to the main house (*villa rustica*) of a farm (Dodt) in the province of Lower Germany.

The third part of the book is dedicated to the cesspit toilets, with case studies from Austria (Petznek), Germany (Dodt), Belgium (Henrotay), the Netherlands (Heirbaut) and Switzerland (Jauch). Whereas the first two of these only deal with a single cesspit, the latter three are all overviews of a number of cesspits found in various contexts within the settlements. These are especially interesting as the different placing of the toilets and their diverse manners of construction give a good indication of the variation that must have been normal among the outhouses in the Northwestern provinces.

The forth part of the book deals with 'portable' toilets; i.e. chamber pots and the like. While Petznek gives a general overview of the different types and uses of chamber pots, Bienert's paper deals with a specific form of pot used in a stool.

The last part on written sources just has one paper by Juntunen, on the use of the word 'stercus' – does it mean human or animal faeces?

It is obvious that at this point, it is impossible to write a definitive history of toilets and toilet-use in Roman times. Much more research is needed to get a clear view of all aspects surrounding human waste removal during the Roman period.

While the basics of the architectural aspects of Roman toilets are better known by now, other aspects have been only touched briefly yet, such as the apotropaic magic used on toilets (Jansen *et al.* 2011: 165-176) or the social implications of multi-seater toilets (Neudecker 1994).

The waste removal processes used in settlements on the countryside and in pre-Roman times are also unknown at the moment.

Therefore, we would like to express out hope that this conference was not the last on this subject in the Northwestern provinces, but just a start for this interesting research topic.

Bibliography

Bradley, M. (ed.) 2012. *Rome, Pollution and Propriety: Dirt, Disease and Hygiene in the Eternal City from Antiquity to Modernity*. Cambridge.

Draycott, J. 2012. *Review on Jansen et al., Roman Toilets: their Archaeology and Cultural History*. Bryn Mawr Classical Review 2012.03.34.

Hobson, B. 2009. *Latrinae et Foricae: Toilets in the Roman World*. London.

Jansen, G. M. C., Koloski-Ostrow, A. O. and Moormann E. M. (eds) 2011. *Roman Toilets: their Archaeology and Cultural History*. BABesch. Supplement, 19. Leuven.

Koloski-Ostrow, A. O. 2015. *The Archaeology of Sanitation in Roman Italy: Toilets, Sewers, and Water Systems*. Studies in the History of Greece and Rome. Chapel Hill, NC.

Magness, J. 2011. *Stone and Dung, Oil and Spit: Jewish Daily Life in the Time of Jesus*. Grand Rapids, MI.

Magness, J. 2012. 'What's the poop on ancient toilets and toilet habits?' *The Journal of Near Eastern Archaeology* 75 (2): 80-87.

Neudecker, R. 1994. *Die Pracht der Latrine: zum Wandel der öffentlichen Bedürfnisanstalten in der kaiserzeitlichen Stadt*. Munich.

Raventós, R X. D. and J-A. Remolà (eds) 2000. *Sordes Urbis: La Eliminación de Residuos en la Ciudad Romana*. Rome.

Scobie, A. 1986. 'Slums, Sanitation and Morality', *Klio*, 68 (2) 399-433.

Thüry, G. E. 2001, *Müll und Marmorsäulen. Siedlungshygiene in der römischen Antike* (Zaberns Bildbände zur Archäologie). Mainz.

Sewers or cesspits? Modern assumptions and Roman preferences

Gemma Jansen

The exploration of Roman toilets, sewers, cesspits or any other kind of waste removal can lead us into a very unglamorous corner of the study of antiquity. The topic either elicits disgust, contempt, or laughter. In the best scenario, we might uncover a secret curiosity among our readers. Within the field itself, research on cesspits is considered somewhat below the study of sewers, which have a higher scholarly status. How can we study this important part of Roman antiquity and still avoid these judgmental feelings?

I have chosen the following approach as an attempt to tackle this problem. First, we investigate the modern assumptions regarding sewers, and discuss what the Romans themselves thought about them. Then, we look for the Roman preferences (sewer or cesspit) by considering the actual situation in Pompeii. What do the archaeological data tell us about Roman system preferences for the removal of liquid waste? A short comparison of the systems in Pompeii with the ones in operation at Herculaneum and Ostia, gives us a better idea of the goals aimed at by the Romans.

Having established the Roman preference, we go back again to the subject of modern assumptions and discuss the question whether a Roman sewer was more hygienic than a Roman cesspit, as this modern idea seems to be one of the reasons for being so condescending about cesspits.

The aim of this paper is to recover – with help of Latin texts **and** archaeological data – the Roman mindset about this subject.

Modern and Roman prejudices about sewers

At first glance, sewers are a rather neutral subject, devoid of value judgments: they form a system of covered underground channels for carrying away liquid waste. As we do not see this underground system of channels and pipes, and as we trust the system to do its job, we are generally quite indifferent to it. The way we look at (Roman) sewers became well established in the recent past.

Modern values coincide with changes in medicine dating from the middle of the 19th century. After several cholera epidemics, the London doctor John Snow discovered in 1854 that this disease was caused by water that was contaminated by faeces containing dangerous bacteria. This is how microbiology became the basis of modern hygiene; it was the group of so-called 'hygienists' who achieved that health care came to be based on microbiology (Horstmanshoff and Stol 2004; Horstmanshoff 2006).

The discovery of bacteria, together with other insights into the appalling living conditions in the poor quarters of towns led to the formulation of rules for public and private health, aimed at preventing disease and maintaining health. The hygienists referred to the excellent hygienic facilities in the classical world, citing Roman sewers and aqueducts as outstanding examples. At the same time, all large European cities were building sewage systems to improve public health. In this light, the sewers of Rome were studied and brought to the fore as the great model for modern Europe. This was the era that the Roman sewers got their labels as 'weapons against disease' and even as 'carriers of civilization' (Van Tilburg 2013).

In my opinion these actions led to our present notion of a sewer as a technical and hygienic step forward (in line with our thinking of continual human progress), implying that cesspits are less valuable because they seem technically uncomplicated and less hygienic.[1] What follows naturally from this is that we are proud of our sewers as tools of improving the state of hygiene, especially when we see this in historical perspective, for example from mediaeval times onwards.

There is more to the sewer: Since we regard human filth and all other waste flowing through sewers as loathsome, we do not want to touch sewer contents, nor sewer workers. Hugo's novel *Les misérables* and the Batman movies (in which the antagonist lives in dark underground sewers) appeal very much to this modern feeling of the sewer as the underbelly of society from which only bad things can come. This negative

---

[1] This has been expressed exemplary by Eschebach (1979: 39): 'Dem Vorhandensein der Stabianer Thermen ist es zuzuschreiben, dass dieser Teil des Stadtgebietes von Pompeji eine aufwendige und für antike Verhältnisse *moderne* Kanalisation erhielt, die, von diesen Thermen ausgehend, das südliche Stadtgebiet der antiken Stadt entwässerte, während sich manche Stadtteile Pompejis auch während der Kaiserzeit noch mit der sonst üblichen, *primitiven* Form der Beseitigung von Abwässern durch Sickergruben (pozzi neri) und Ableitung des Wassers auf die Fahrbahnen gepflasterter Straßen begnügen mussten.'

Figure 1. Rome, Cloaca Maxima, dr. Heinrich Bauer in foreground (photo G. Jansen)

undertone can be illustrated with a rather recent example. In June 2008, there was a proposal in San Francisco to rename a sewage plant after the departing president George W. Bush (New York Times June 25, 2008). This was meant as an insult, and even though everybody would agree that a sewage plant in itself is a very good thing, everyone could also understand the insult. In short: we believe we are neutral towards sewers. When, however, we take a closer look, this clear picture blurs a little: while we highly appreciate their technical framework, their contents actually horrify us.

It seems that the Romans had this ambiguous attitude as well. They were proud of their sewers and especially the Cloaca Maxima as large monumental and impressive structures (Figure 1, for information on the Cloaca Maxima see Bauer 1993-2000, Bianchi 2015 and Placidi *et al* 2013).

One can still feel the unconcealed pride with which Pliny (NH 36: 104-106) writes about the sewers in his enumeration of the wonders of Rome 'But at the same time elderly men still admired [...] the city sewers, the most noteworthy achievement of all, seeing that hills were tunnelled and Rome [...] became a 'hanging' city, beneath which men travelled in boats during Marcus Agrippa's term as *aedile* after his consulship. [...] In the streets above, massive blocks of stones are dragged along, and yet the tunnels do not cave in. They are pounded by falling buildings, which collapse of their own accord or are brought crashing to the ground by fire. The ground is shaken by earth tremors; but in spite of all, for 700 years from the time of Tarquinius Priscus, the channels have remained well-nigh impregnable.' (Loeb translation by Eichholz 1971).

Dionysius of Halicarnasus (3, 5) also praises the sewers as 'a wonderful work exceeding all description. Indeed in my opinion the three most magnificent works of Rome, in which the greatness of the empire is best seen, are the aqueducts, the paved roads and the construction of the sewers.' (Loeb translation by Cary 1961). Though the Romans praised the Cloaca Maxima and other sewers, at the same time they despised their contents. Livy (1.56.2), for example, speaks highly in one sentence about the sewer as one of 'the two works for which the new splendour of these days has scarcely been able to produce match', but at the same time refers to it as the 'receptacle (dustbin) for all the off scouring of the city' ('*receptaculum omnium purgamentorum urbis*' Loeb translation by Foster 1967).

This also finds expression in the habit of humiliating opponents by disposing them like rubbish in the sewers. I refer to the death of the emperor Heliogabalus, who tried to hide himself in a public toilet and was dragged out of it by his soldiers. After they killed him, they wanted to dispose of him via a sewer shaft. As the shaft was too small, however, they eventually threw him in the river

Tiber: '[..] the body was dragged through the streets, and the soldiers further *insulted* it by thrusting it into a sewer' (*HA* Antonius Elagabalus 17, 1-2 and 23, 7, Loeb translation by Magie 1967). The other insulting aspect involved here is of course the fact that the body remains unburied.

During civil wars or riots, murdered senators and common people were also thrown into the sewers to humiliate them (See for example Livy 1.14, Cicero, Pro Sestio 15.77 and HA the three Gordians 13, 8). Even statues were thrown into sewers to humiliate the person they represented. Suetonius (Nero 24.1) tells us that the emperor Nero wanted to wash out the fame of other winners at a contest by throwing their statues in the toilet (sewer). The 2005 find of a colossal head of the emperor Constantine in a sewer in Rome seems to underline this way of thinking.[2]

We can conclude that modern people are proud of sewer structures, but despise its contents and the same is true of the Romans. We look at Roman ambiguity with our own ambiguous view. In addition, we relied to a great extent upon the examples of Roman sewers when we started to build sewers in the modern world. And this led to the common opinion that the Romans were sophisticated sewer builders and that sewers were an essential part of their civilization.

The latter is an anachronistic misinterpretation that can lead to incorrect conclusions or reconstructions of the Roman world. For instance, in the Roman town of Pompeii only a few sewer lines have been found. According to some researchers (Eschenbach 1979; Mygind 1921), this implies that the town was not really Roman and therefore not yet civilized. Another scholar (Strell 1913) explains this situation by saying that – at the moment – only part of the sewer system has been excavated: after all, Pompeii is a Roman town, *ergo* it must have a complete sewer network. Both hypotheses do not tell us anything about Roman sewers, but testify to our firm idea of the sewer as representative of Roman civilization. The hypotheses also highlight that other workable methods of waste disposal, such as cesspits and disposal along sloping streets, have been overlooked. My own research (Jansen 2000 and 2002) indicates that, in Pompeii, the latter methods were in fact preferred, quite calculatingly, over an expensive sewer system that, moreover, needed a lot of maintenance.

Roman preferences, the example of Pompeii

The Romans had several options at hand to dispose of rubbish, including urine and excreta. The example of Pompeii can illustrate this well. The slope on which the city was built facilitated an easy discharge of rain and wastewater, and the permeable subsoil was ideal for dealing with the urine and faeces of the city's 10,000 inhabitants. Arthur (1993: 194) gives a rough calculation of the amounts of waste that had to be disposed of yearly: 3 650,000 litres urine and 730,000 kilo's of faeces. The amounts of urine necessary for the fuller's process were little and will not have lowered these figures substantially (see Flohr and Wilson 2011: 147-157). What happened to other waste materials is not clear though rubbish pits have been found all over the city.[3]

Urine and faeces were collected at public and private toilets. Most of the 10 public toilets were located in large baths and were automatically flushed once a day or continually (Manderscheid 2009: 29-31; Van Vaerenbergh 2011: 78-86). In contrast, private toilets inside a house were built in such a way that they could be flushed with a bucket of water. A small plateau of tiles was built in front of the toilet seat for this purpose: the tiles slanted slightly towards the drain underneath the seat (Figure 2). Some toilets are constructed differently. These are built in a niche into the wall and the seat is right above a vertical drainpipe (Figure 3). These occur on upper floors (Trusler and Hobson 2007) or at other places where there was little water available for flushing. A closer look at the location of toilets shows that many are situated in gardens or rooms near the streets. Other favourite places are underneath stairs and in (or near) kitchens.

In the past, I investigated more than 200 of these private toilets (Jansen 1997), while Barry Hobson (2009 and 2011) has found many more in the meantime. We can safely state that a Pompeian house had at least one toilet; even the rented apartments on the upper floors were provided with one. For the Pompeians, a toilet was a standard facility.

These toilets in houses were connected to cesspits built according to the idea of a 'soakaway': liquid material seeped into the ground and solid material remained in the pit. This result was achieved by constructing the pit's wall in such a way that the liquids could get out. Of the hundreds of toilets that have been detected in Pompeii so far, only a few have had their cesspit excavated completely. One of the reasons that these

---

[2] In first reports on Internet and in newspapers the find was interpreted as a humiliation of Constantine. This thought is also reflected in the official museum text in the Mercato di traiano – museo dei fori imperiali, where the head is presently exhibited. In the later articles (Hannestad 2007: 102-105; La Rocca and Zanker 2007) this prior assumption is not mentioned again, nor is it revoked.

[3] Rubbish pits have been excavated in *atria* (Dickmann and Pirson 2002: 306-307), in gardens (Robinson 2005) and in streets (Berg 2008). Several excavators complain that it is hard to distinguish between a rubbish pit and a cesspit (for example, Arthur 1986: 31, 35, 38; Arthur 1993: 195; Eschebach 1982; Robinson 2005. See also Heirbaut in this volume). For interesting notions on rubbish in Pompeii see Ciaraldi and Richardson (1999) who studied the waste deposits of two houses. They concluded from the relatively small amounts of domestic debris found and from the lack of larger animal bones that some formalized kind of rubbish collection must have existed in Pompeii.

Figure 2. Pompeii, Casa di Apollo VI 2, 15.22, (hand) flush toilet in kitchen (photo G. Jansen)

Figure 3. Pompeii, house V 1, 30, drop toilet on first floor (photo G. Jansen)

have not been excavated down to the bottom is that it is often dangerous to work in the Pompeian underground. This is due to the presence of *mofetà*, the poisonous gas, but also to the danger of collapse. Among others, Hobson (2009: 47, 63) refers to this problem. The known number of cesspits has increased by up to 51 examples in recent years as many of the international teams working in Pompeii have excavated large numbers of them (see Guzzo and Guidobaldi 2005; 2008). Many been excavated by Berg (2008: 363-375). For the location and literature of these latrines, see Table.

From this sample we can state that there was no standard cesspit in Pompeii: depth and shape differ enormously. Most of the cesspits are round (Figures 4 and 5), though they can be square, elliptical or lozenge-shaped. Only seven have been excavated completely and their depths vary from 2.5 m up to 11.2 m.[4] A few are just holes dug in the ground, though most have some reinforcement at the top: the upper part is mostly in stone (with or without mortar) and the lower part is dug in the underground. This construction ensures that the liquids can seep out of the pit. Three of the cesspits deviate from this spectrum, as they were re-used existing underground structures like cisterns, cellars or wells.[5]

---
[4] The depths are 2.5 m, 2.8 m, 5.4 m, 5.4 m, 6 m, 6.3 m and 11.2 m. From this small sample one can conclude that the 11.2 m pit is an exception.
[5] Cistern reused as cesspit in Casa degli Amorini Dorati (VI 16, 7.38), see Seiler 1992: 52-53. Cellar reused as cesspit in Casa degli Capitelli Colorati (VII 4, 31-33, 50-51), see Descoedres and Sear 1987: 12 fig.

Figure 4. Pompeii, Casa del Meandro (I 10, 4.15), location of cesspit outside the house (small window is window of the toilet room) (photo G. Jansen)

To facilitate emptying, the pits usually had a second opening.[6] Of the 51 cesspits known, five have their second openings in the toilet room itself, five in the nearby kitchen, five in the garden (Figure 6) and 15 on the sidewalk of a nearby street. The other cesspits either have their openings elsewhere in the house or the excavator did not mention the location.

Some openings still have their original covers: on sidewalks, this mostly is a stone slab (Figure 7), and in houses, a stone lid with a metal ring (Casa delle Vestali (VI 1, 6-8, 24-26), house VI 2, 27 and house VII 12, 11) (Figure 8). There is also evidence of wooden covers: Hobson (2009: 63-64) describes the find of nails in the upper part of a cesspit of Casa delle Vestali (VI 1, 6-8, 24-26), and Berg (2008: 365) has found part of the burned wooden covering. Several cesspits have been found open and one is inclined to think that they were being emptied at the moment of the eruption of Mount Vesuvius. In one case, cesspit contents have been found on the street next to three open cesspits (Berg 2005: 200). The cesspit of house VIII 4, 40-40a (room 1) was also open and Dickman and Pirson (2002: 288-289, 303-304) suggest this was either for emptying or for repairs.

We have no definite evidence yet about where the contents were brought under normal circumstances. It seems safe to assume though that they were used on the fields outside the city as many Roman authors refer to this practice of manuring. Arthur (1993: 195) refers to

Figure 5. Pompeii, Casa del Meandro (I 10, 4.15), detail of cesspit (photo G. Jansen)

1 and 17, see also Sear 2006: 181-184. Well reused as cesspit in Casa della Fontana Picola (VI 8, 23-24), see Fröhlich 1996: 70-71 and Abb. 459-460.

[6] In contrast, one of the cesspits excavated by Berg (2008: 366-368) was closed at the top. To empty it, the top would have to have been smashed. Accordingly, this one was never emptied.

Figure 6. Pompeii, Casa delle Nozze d'Argento (V 2, i.e.25) cesspit opening in garden at the back of the toilet room (photo G. Jansen)

Figure 7. Pompeii, street pavement, stone cesspit cover (photo G. Jansen)

Figure 8. Pompeii, house VI 2, 27, cesspit cover with metal ring in kitchen (photo B. Hobson)

Jashemski 1979 (without page number), suggesting that she has evidence of manuring the fields outside the city with the contents of cesspits, but there is no evidence in her book. Parslow (2000: 208) also claims – without citing evidence – that the sewage water that was led to the soakaway in the back yard of the Casa di Iulia Felix (II 4) was used for the garden. Robinson (2005: 113, 118) came across a shallow pit (no 411) in the garden of house I 9, 11-12 which he regards as the remains of a temporary toilet; the other possibility, however, is that the content of a cesspit were deposited here. This would then imply that not all night soil was used to fertilize gardens and fields, as described in ancient texts (See also Flohr and Wilson 2011: 147-157).

Now that we know that excreta and urine (together with wastewater from kitchens etc.) were discharged into cesspits, we have to ask the question where other liquid waste, such as rainwater and the overflow from street fountains, was drained. Pompeii was fortunate in this aspect, too. Because of the slope of the streets, rainwater automatically flowed downwards and left the city through the gates. At such times, the high sidewalks and stepping-stones proved to be no superfluous luxury. The Pompeians did not just sit back and watch the water leave their city, but tried to manipulate it, to prevent flooding of some streets and to reduce the flow in others. One method of manipulating the water was to raise the pavement subtly and thus force the water to change direction (see also Poehler 2012).

Looking at these installations, one could become convinced that Pompeii did not need a sewer – its excreta were disposed of in cesspits and the sloping streets carried its rainwater away. But Pompeii had a sewer anyway. It was not a complete sewer network, but consisted of different and separate branches, built to solve specific problems. One system of branches had been build to drain the Forum (Arthur 1986: 36-37, Pl IVb; Cozzi 1900: 592-594, Kockel and Flecker 2008: 286, 287, 289, 292, Abb. 11/12, 15, 16, 17, 19/20). Most of the other branches were connected to buildings where large amounts of water had to be disposed of, like the baths. For example the large toilet of the Stabian baths marks the beginning of a sewer line.

As soon as the branches of a sewer were laid out, they started to be used to dispose other waste as well. In places where there was a lot of rainwater in the streets, a connection was made to the sewer to remove the excess. Also, toilets could now be connected to the sewer.[7] This was what happened with the sewer that started at the Stabian baths. Several toilets in the insula south of the baths were connected to the sewer: these were placed right above the sewer line. Both the sewer contents as well as the wastewater drained by the streets must have ended up in the Sarno river.

To sum up, one could say that because of the porous subsoil, excreta could be collected in cesspits. Due to the sloping streets and with only little extra manipulation, rainwater and wastewater could be directed out of the city. A small number of sewer branches were built to cope with specific problems: for instance, to lead away large amounts of wastewater from baths. In some places rainwater was led into the sewer, and in other cases the toilets discharged into it. Most importantly, one can conclude that, at Pompeii, different kinds of waste were disposed of by means of different methods, and that sewers played only a minor role in this system.

Other examples

A comparison of the situation at Pompeii to Herculaneum and Ostia can show whether the provisions chosen for Pompeii are unique. Herculaneum was also built on a lava platform with a steep slope; though in this case the underground was *not* porous, but hard and compact (Camardo 2007; Camardo and Notomista 2015; Jansen 1991; 2000a; 2002).

So here, rainwater could easily be removed through the sloping streets, but its compact tufa subsoil caused cesspits to be filled up quickly – though Maiuri (1958) keeps referring to them as *pozzi assorbenti*.

This system was not functioning well and one can imagine that cesspits had to be emptied often. The evidence is meagre. Of the 88 toilets identified so far only three have a cesspit excavated: one round cesspit in house IV, 18 room 4 (Camardo 2007: 177 and fig 10), one in the service corridor of the Central Baths and one in the north corner of the premise of the Augustali (Camardo and Notomista 2015: 58-59) We know of another that it has been emptied: on a column in the peristyle of the Casa del Salone Nero (VI 11-13) an inscription could be read at the moment of excavation: *exemta ste(r)cora a(assibus) XI* ('the cesspit has been emptied for eleven as' *CIL*. IV. 10606). In this house, a toilet for two persons had been discovered; the cesspit has not yet been excavated.

The inconvenience of cesspits that were quickly full was tackled provisionally in Herculaneum by constructing toilets in such a way that they did not need much flushing: niche toilets were preferred. Later, installing a systematic sewer network solved the problem. By and by, toilets were connected to this system and the inhabitants of Herculaneum could now install flush toilets. But not all choose to build flush toilets. Where water was scarce, drop toilets were preferred, for example in Insula orientalis II.[8] The sewage from both the streets and the sewer was drained towards the beach and the sea.

At Ostia, things were rather different. The circumstances were not so favourable, due to its position in the Tiber delta. The city was built on flat sand dunes with depressions in the middle of the town, so it was impossible to discharge rainwater and wastewater out of the town via the streets. Besides, due to the closeness of the river, the ground water table was at only 2 to 3 meters below walking level. This made cesspits not a logical choice; indeed, from the very beginning, an extensive sewer network was laid out (Figure 9).

---

[7] But I know of only one private toilet that is visibly connected to a sewer at the moment: the toilet of VIII 4, 8 together with a drainpipe from upstairs. Dickmann and Pirson (2000: 454, 467; 2002: 291) discovered that the toilet of the Casa dei Posthumii (VIII 4, 4, 49) and the toilet at the back of shop VIII 4, 35 were connected to the sewer. Several earlier excavators and researchers mention toilets connected to a sewer: toilet of VI 17, 1-4 (Piranesi 1804: plate II and XXX), Casa di Giuseppe II (VIII, 2 38-39) (Mau 1887: 121), house IX 7, 12-13 (Mau 1889: 105), house IX 7, 15-20 (Mau 1890: 238), Casa di Pansa (VI 6, 1.8.11-13) (Strell 1913: 99-103), Casa di Salustio (VI 2, 3-5.30) (Strell 1913: 99-103). I tried to verify these last references on site, but none of them is visible at the moment. Eschebach (1982) also describes and shows on Tafel 89, 3-4 a down pipe ending in the small sewer underneath the street. The down pipe is still visible, the sewer he is referring to, however, is not.

[8] Recent research has questioned whether the 'sewer' underneath this insula was a sewer, a large cesspit or at least a malfunctioning sewer (Camardo *et al.* 2006; Camardo 2007; 2011).

Figure 9. Ostia, sewer and house connections in the sidewalls (photo G. Jansen)

This network, which had been repaired and extended ever since, has remained almost completely intact underneath the city and can be entered through the many sewer covers (Figure 10, see Jansen 1995; Jansen 2000a; Jansen 2002). More than 50 of these covers can still be seen in the Ostian streets (Figure 11).

Having looked at the various situations in the three towns discussed, it can be concluded that the inhabitants preferred to get rid of their rainwater, wastewater and the overflow water of fountains by way of the streets. Only when circumstances made it impossible – when the amount of water was too large, for instance, or when a city did not have sloping streets – the waste flows were led away by sewers. Excreta were preferably collected in cesspits, unless the subsoil was not suitable for such pits. In that case, the excreta were disposed of through a sewer.

Furthermore, we can state that for each city, the environmental circumstances were different (subsoil, gradient of the city landscape etc.) and as the Romans opted for made-to-measure solutions, their choices were different. There was no standard solution and even when the same system was installed, it was executed differently in every town, according to the circumstances and materials available. The inhabitants preferred to install small-scale provisions that required low maintenance, like cesspits. Only when these were not effective, they were obliged to install large infrastructural works, like a network of sewers. This was a much more expensive installation and it required municipal intervention to organize the cleaning and upkeep.

We can also draw a conclusion about the type of toilet the Romans preferred: toilets connected to sewers or toilets draining into cesspits. We saw from our description of the toilets at Pompeii that group toilets were flushed automatically; private toilets could be hand flushed and drop toilets did not need much flush water. The types of toilets were chosen according to functionality. Where large crowds had to be served and much water was at hand, automatically flushed toilets were installed. In houses, the type of installation chosen also depended on the amount of water available: with much water available a flush toilet was convenient, in apartments on upper floors and in shops, with not much water at hand, drop toilets were more efficient. All types of toilets could – in theory – be connected to a cesspit or a sewer, though the automatically flushed toilets seldom had a connection to a cesspit. Again, Romans seemed to be pragmatic and had no absolute preference.

Were Roman sewers really that hygienic?

Once more, we go back to the labyrinth of assumptions and we have to investigate the question whether the building of sewers made the Roman world healthier and more hygienic. When dealing with Roman sewers,

Figure 10. Ostia, manhole into sewer in the street of the Baths of Mistras (photo G. Jansen)

Figure 11. Ostia, sewer cover in the Decumanus Maximus (photo G. Jansen)

we think they look like our sewers and we think they were installed with the same intention as ours. This is also true for Roman aqueducts and Roman toilets: it is hard to let go of our own modern concepts (see Jansen 2000b). It is true that our modern world definitely became more hygienic by laying out sewers. But even in this case we cannot look at sewers as an isolated phenomenon. It was not the mere installation of sewers that improved our modern world; there were a host of things that helped. First, more important than laying out sewers was the discovery of bacteria and then the awareness that there was a link between these bacteria and diseases. Secondly, building sewers was just **one** of the measures taken to improve hygienic conditions. This is the opposite from the situation in antiquity, where no one was aware of bacteria, nor of the diseases related to them. As a result, there were no measures taken in this field.

In the Roman world, building sewers was a rather practical thing: to get rid of dirt when there was no other or easier option available. There is no reference in any ancient text that sewers were being build for hygienic purposes, there is only a vague notion that keeping the sewers clean was improving the 'health' of the citizens (see below).

We must realize, that the Roman world was far from hygienic (Mitchell 2015). Even with sewers available, the streets remained as dirty as they were before (full of animal dung and all kinds of organic waste), and toilets remained positioned in kitchens. As Roman toilets had no stench trap or swan neck, toilets were directly connected to the sewer or cesspit underneath. So rats and other animals and insects could enter the kitchen and spread all kinds of diseases there.[9] Consequently, it would be fair to state, that the installation of toilets and sewers did not improve hygiene (Jansen 2011).

Moreover, Roman sewers were not self-cleaning and their shape did not activate fluids and more solid materials to flow away easily. This is due to the difference in slope and shape: the modern sewer is V-shaped and the Roman U-shaped (see Figure 9). Every now and then, men had to go down to free the sewers from deposits and take away blockages (Dionysius of Halicarnassus 3 67, 5). This is the reason for Ostia's many manholes giving access to the sewer system. In the Digest (43.23.1) it is mentioned that the sewers must be cleaned and maintained to keep the citizens in health (*salubritas*) and the system safe. However, the precise meaning of 'health' here is not so clear.

Most sewers were small and dark, and working in it must have caused many difficulties. A study into the nineteenth century Parisian sewers (Reid 1991: 149, 153-155) highlights the dangers of working in the sewers and there is no reason to assume that they differ from the dangers in the Roman sewers: Serious falls were surely quite common in the small, dark and slippery sewers, and the constant cool dampness could bring about hernias and arthritis. The fauna of the sewer was often dangerous, too: spiders and poisonous centipedes could cause nasty bites that could easily turn into infections, as the sewer environment was full of bacteria. The rotting material in the sewers attracted swarms of flies; rats were dangerous when cornered: they jumped and bit into hands and faces and, at worst, they infected a person with a fatal virus. In addition, other dangers were present, such as the risk of drowning in case of the sewers quickly filling up during a downpour or the risk of sudden death by inhaling poisonous gases.

In fact, during a health research of the sewer workers of Paris, carried out around 1900, it was discovered that this work was so unhealthy that one third of the working crew died within ten years of starting to do this underground work.

The Romans regarded working in the sewers as risky as 'walking on a slack-rope', but they do not specify the dangers and so we do not know whether they are referring to the abovementioned perils (Digest 19.1.54). Whatever the dangers are, they caused the emperor Trajan to advise Pliny in a letter (10, 32) to use former prisoners for 'cleaning public baths and sewers, repairing streets and highways' which he regards as 'the usual employment for men of this type' (Loeb translation by Radice 1969).

Though Roman sewers might have improved the general health of the population in the city by evacuating all kinds of (fluid) rubbish, at the same time they created unhealthy situations in the areas around the toilets (especially in kitchens). The people who had to clean the sewers worked in extremely unhealthy conditions.

Conclusion

In this paper I have tried to show that while studying sewers and cesspits, we find ourselves in a mental labyrinth. We have all kinds of (positive) feelings regarding these installations and (negative ones about) their contents; and so did the Romans, who were proud of sewer construction, but despised the filling. But there is a very big difference between the Romans and us: we regard sewers as bringing health, hygiene and civilization, whereas the Roman mindset about sewers (and cesspits) is more practical. These were not built with the aim to improve public health or hygiene. They were constructed to dispose of urine and faeces in an easy way. The Romans choose to install a sewer when a cesspit did not work or did not work well. Digging a cesspit was their first choice. It was low cost, had low maintenance, and did not need municipal intervention.

The choices taken by Romans cannot be detected by studying some sewer lines or a group of cesspits in isolation, but only by investigating the complete ensemble of systems for (liquid) waste disposal in operation within one city. We have to look with new eyes at Roman sewers and cesspits. We have a long way to go, and this paper might serve as a beginning.

Acknowledgement

I would like to thank Ann Koloski-Ostrow for much appreciated discussions on this topic and her kind revision of this text.

---

[9] Sewer rat (*sorex cloaca*) is used as an abusive word in a poem (*Anthologia Latina* 196.1). Rats are not only known from texts, but more and more from archaeological contexts. Arthur (1986: 35, 41; 1993: 197) excavated a 2nd century BC rat skeleton at Pompeii. Thüry (2001) provides more examples.

Bibliography

Amorso, A. 2005. Analisi strategrafica e prime proposte di ricostruzione dell'insula VII 10 di Pompei. In P. G. Guzzo and M. P. Guidobaldi (eds) *Nuove ricerche archeologiche a Pompei ed Ercolano, Atti del convengo internazionale, Roma 28-30 Novembre 2002*: 36-59. Napoli.

Arthur, P. 1986. Problems of the urbanization of Pompeii: Excavations 1980-1981. *The Antiquaries Journal* LXVI 1: 29-44.

Arthur, P. 1993. Le città vesuviane: problemi e prospettive nello studio dell'ecologia umana nell'antichità. In L. Franchi dell'Orto (ed.) *Ercolano 1738-1988: 250 anni di ricerca archaologica, Atti del Convengno Internazionale Ravello-Ercolano-Napoli-Pompei 30 ottobre-5 novembre 1988*: 193-199. Roma.

Bauer, H. 1993-2000. Cloaca Maxima. In E. M. Steinby (ed.) *Topographicum Urbis Romae*, 288-290. Rome.

Berg, R. 2005. Saggi archaeologici nell' insula dei Casti Amanti. In P. G. Guzzo and M. P. Guidobaldi (eds) *Nuove ricerche archeologiche a Pompei ed Ercolano, Atti del convengo internazionale, Roma 28-30 Novembre 2002*: 200-215. Napoli.

Berg, R. 2008. Saggi stratigrafici nei vicoli a est e a ovest dell' insula dei casti Amanti (IX 12). Materiali e fasi. In P. G. Guzzo and M. P. Guidobaldi (eds) *Nuove ricerche archeologiche nell'area vesuviana (scavi 2003-2006. Atti del Convegno Internazionale, Roma 1-3 febraio 2007*: 363-375. Napoli.

Bianchi, E (ed.) 2015. *Cloaca maxima. E i sistemi fognari di Roma dall'antichità ad oggi*. Roma.

Camardo D., Martelli Castaldi M. and Thompson J. 2006. Water Supply and Drainage at Herculaneum. In G. Wiplinger (ed.) *Cura Aquarum in Ephesos*, BABesch suppl. 12/ Sonderschrift ÖJh 42: 183-191.

Camardo, D. 2007. Ercolano: la gestione dell acque in una città romana. *Oebalus, Studi sulla Campania nell' Antichità* 2: 176-185.

Camardo, D. 2011. Ercolano: la riconstruzione dei sistemi fognari di un'antica città. In G. Jansen, A. Koloski-Ostrow and E. Moormann (eds), *Roman toilets. Their Archaeology and Cultural History, BABesch supplement* 19: 90-91.

Camardo, D. and M. Notomista 2015. Le latrine di Herculanuem. Studio dei sistemi igienici di una città romana, *Vesuviana* 7: 55-190.

Cerulli Irelli, G. 1977. Un' officina di lucerne fittili a Pompei. In *L'instrumentum domesticum di Ercolano e Pompei nella prima età imperiale*: 53-72. Roma.

Ciaraldi, M. and Richardson J. 2000. Food, ritual and rubbish in the making of Pompeii. In G. Fincham at al (ed.) *TRAC 99 Proceedings of the ninth annual Theoretical Roman archaeology Conference, Congress Durham 1999*: 74-82.

Cozzi, S. 1900. La fognatura di Pompei. *Notizie degli Scavi*: 587-599.

Descoedres, J. P. and Sear F. 1987. The Australian expedition to Pompeii. *Rivista di Studi Pompeiani* I: 11-36.

Dickmann, J. A. and Pirson F. 2000. Die casa dei Postumii VIII 4, 4-49 in Pompeji und ihre *insula*, Bericht über die 3. Kampagne 1999. *Mitteilungen des Deutschen Archeologischen Instituts, Römische Abteilung Band* 107: 451-467.

Dickmann, J. A. and Pirson F. 2002. Die casa dei Postumii in Pompeji und ihre Insula, *Mitteilungen des Deutschen Archeologischen Instituts, Römische Abteilung Band* 109: 243-316.

Ellis, S. and Devore G. 2008. The Fourth season of Excavations at VIII.7.1-15 and the Porta Stabia at Pompeii: preliminary report. *Fasti On Line Documents & Research*. No. 146.

Eschebach, H. 1979. *Die Stabianer Thermen in Pompeji*. Berlin.

Eschebach, H. 1982. Die Casa di Ganimede in Pompeji VII 13, 4. *Mitteilungen des Deutschen Archeologischen Instituts, Römische Abteilung, Band* 89: 228-436.

Flohr, M. and Wilson A. 2011. The economy of ordure. In G. Jansen, A. Koloski-Ostrow and E. Moormann (eds), *Roman toilets. Their Archaeology and Cultural History*, (BABesch supplement 19): 147-157.

Guzzo, P. G. and Guidobaldi M. P. (eds) 2005. *Nuove ricerche archeologiche a Pompei ed Ercolano, Atti del convengo internazionale, Roma 28-30 Novembre 2002*. Napoli.

Guzzo, P. G. and Guidobaldi M. P. (eds) 2008. *Nuove ricerche archeologiche nell'area vesuviana (scavi 2003-2006. Atti del Convengo Internazionale, Roma 1-3 febraio 2007*. Napoli.

Fröhlich, Th. 1996. *Casa della Fontana piccola VI 8, 23.24. Häuser in Pompeji* 8. München.

Hannestad, N. 2007. Die Porträtskulptur zur Zeit Konstantins der Grossen. In A. Demandt and J. Engelmann, *Konstantin der Grosse – Ausstellungskatalog*: 96-115. Mainz.

Hobson, B. 2009. *Latrinae et foricae*. London.

Hobson, B. 2011. The Location of Latrines in Roman Houses at Pompeii. In G. Jansen, A. Koloski-Ostrow and E. Moormann (eds), *Roman toilets. Their Archaeology and Cultural History* (BABesch supplement 19): 123-130.

Horstmanshoff, H. and Stol M. (eds) 2004. *Magic and Rationality in Ancient Near Eastern and Graeco Roman Medicine*. Leiden/Boston.

Horstmanshoff, H. 2006. *Patiënten zien. Patiënten in de antieke geneeskunde*. Leiden.

Jansen, G. 1991. Water Systems and Sanitation in the Houses of Herculaneum. *Mededelingen van het Nederlands Instituut te Rome* 50: 144-166.

Jansen, G. 1995. Die Wasserversorgung und Kanalisation in Ostia Antica; Die ersten Ergebnisse. *Mitteilungsheft der Frontinus-Gesellschaft* 19: 111-123.

Jansen, G. 1997. Private toilets at Pompeii: appearance and operation. In S. E. Bon and R. Jones (eds) *Sequence and Space in Pompeii*: 121-134. Oxford.

Jansen, G. 2000a Systems for the disposal of waste and excreta in Roman cities. The situation at Pompeii, Herculaneum and Ostia. In X. Dupré Raventos and J.-A. Remolà (eds) *Sordes Urbis, La eliminación de residuos en la ciudad romana*: 37-49. Roma.

Jansen, G. 2000b Studying Roman hygiene: the battle between the 'optimists' and the 'pessimists'. In G. Jansen (ed.) *Cura Aquarum in Sicilia, Proceedings of the 10th International Congress on the History of Water Management and Hydraulic Engineering in the Mediterranean Region 1998* (BABesch supplement 6): 275-279. Leiden.

Jansen, G. 2002. *Water in de Romeinse stad. Pompeji - Herculaneum - Ostia*. Leuven.

Jansen, G. 2011. Toilets and Health. In G. Jansen, A. Koloski-Ostrow and E. Moormann (eds), *Roman toilets. Their Archaeology and Cultural History*, (BABesch supplement 19): 157-162.

Jashemski, W. 1979. *The Gardens of Pompeii, Herculaneum and the villas destroyed by Vesuvius*. New York.

Jones, R. and Robinson D. 2005. The economic development of the commercial triangle VI.1.14-18, 20-21. In P. G. Guzzo and M. P. Guidobaldi (eds) *Nuove ricerche archeologiche a Pompei ed Ercolano, Atti del convengo internazionale, Roma 28-30 Novembre 2002*: 270-277. Napoli.

Kockel, V. and Flecker M. 2008. Forschungen im Südteil des Forums von Pompeji. Ein Vorbericht über die Arbeitskampagnen 2007 und 2008. *Mitteilungen des Deutschen Archeologischen Instituts, Römische Abteilung* 114: 271-303.

La Rocca, E. and Zanker P. 2007. Il ritratto colossali di Constantino dal Foro di Traiano. In A. di Leone, D. Palombi and S. Walker (eds) *Res Bene Gestae: ricerche di storia urbana su Roma antica in onore di Eva Margareta Steinby*: 145-168. Roma.

Ling, R. 1997. *The Insula of the Menander at Pompeii*. Oxford.

Maiuri, A. 1958. *Ercolano: I Nouvi Scavi I 1927-1958*. Roma.

Manderscheid H. 2009. *Dulcissima Aequora. Wasserbewirtschaftung und Hydrotechnik der Terme Suburbane in Pompeii* (BABesch supplement 13). Leiden.

Mau, A. 1887. Scavi di Pompei 1885-86, Reg. 8 ins. 2 38.39: Casa di Giuseppe II, *Mitteilungen des Deutschen Archeologischen Instituts, Römische Abteilung* 2: 110-138.

Mau, A. 1889. Scavi di Pompei 1886-88, Insula IX 7 (continuazione), *Mitteilungen des Deutschen Archeologischen Instituts, Römische Abteilung* 4: 101-125.

Mau, A. 1890. Scavi di Pompei Insula IX, 7 *Mitteilungen des Deutschen Archeologischen Instituts, Römische Abteilung* 5: 228-284.

Mitchell, P. D. 2015. Human parasites in the Roman world: health consequences of considering an empire, *Parasitology* doi: 10.1017/S0031182015001651.

Mygind, H. 1921. Hygienische Verhältnisse im alten Pompeji. *Janus* 25: 251-281.

Parslow, Ch. 2000. The hydraulic system in the balneum venerium et nongentum of the Praedia Iulia Felicis in Pompeii. In G. Jansen (ed.) *Cura Aquarum in Sicilia, Proceedings of the tenth International Congress on the History of Water management and Hydraulic Engineering in the Mediterranean Region, Syracuse 1998* (BABesch supplement 6): 201-209.

Peters, W. J. Th. (ed.) 1993. *La casa di Marcus Lucretius Fronto a Pompei e sue pitture*. Amsterdam.

Piranesi, F. 1804. *Antiquités de la Grande Grèce. Antiquités de Pompéia I-II*. Paris.

Placidi, M., Valenti S. and Weustink I. 2013. Die Abwasserleitungen Roms. In Frontinus Gesellschaft (ed.), *Iulii Frontini De Aquaeductu urbis Romae, Die Wasserversorgung der Stadt Rom*: 253-266. München.

Poehler, E. 2012. The drainage system at Pompeii: mechanisms, operation and design. *Journal of Roman Archaeology* 25: 95-120.

Reid, D. 1991. *Paris Sewers and Sewermen, Realities and Representations*. Cambridge.

Robinson, M. 2005. Fosse, picolle fosse a peristili a Pompei. *Nuove ricerche archeologiche a Pompei ed Ercolano, Atti del convengo internazionale, Roma 28-30 Novembre 2002*: 109-119. Napoli.

Sear, F. 2004. Cisterns, drainage and lavatories in Pompeian houses, Casa del Granduca (VII.4.56). *Papers of the British School at Rome*: 125-166.

Sear, F. 2006. Cisterns, drainage and lavatories in Pompeian houses II, Casa dei capitelli colorati (VII.4.51), Casa della caccia antica (VII.4.48) and Casa dei capitelli figurati (VII.4.57). *Papers of the British School at Rome*: 163-201.

Seiler F. 1992. Casa degli Amorini d'orati (VI 16,7). *Häuser in Pompeji* 5. München.

Staub Gierow, M. 2008. Some results from the last years' field work in Casa degli Epigrammi Greci (V 1, 18.11-12). In P. G. Guzzo and M. P. Guidobaldi (eds) *Nuove ricerche archeologiche nell'area vesuviana (scavi 2003-2006. Atti del Convengo Internazionale, Roma 1-3 febraio 2007*: 93-108. Napoli.

Strell, M. 1913. *Die Abwasserfrage in ihrer geschichtlichen Entwicklung von den ältesten Zeiten bis zur Gegenwart*. Leipzig.

Strocka, V. M. 1984. La Casa del Principe di Napoli (VII 5, 7.8). *Häuser in Pompeji* 1. Tübingen.

Thüry, G. 2001. Ratte. *Der Neue Pauly. Enzyklopädie der Antike*. Band 10, 785-786. Stuttgart-Weimar.

Trusler, A. K. and Hobson, B. 2017. Down pipes and upper story latrines in Pompeii *Journal of Archaeological Science: Reports* 13, 652-665.

Van Tilburg, C. 2013. A 'Healthy Mistake': The Excrement Problem from Ancient Greece to Nineteenth

Century Holland. In A. Karenberg, D. Groß and M. Schmidt (eds) *Forschungen zur Medizingeschichte. Beiträge des 'Rheinischen Kreises der Medizinhistoriker'*: 103-117. Kassel.

Van Vaerenbergh, J. 2011. Flush water for toilets in and near baths. In G. Jansen, A. Koloski-Ostrow and E. Moormann (eds), *Roman toilets. Their Archaeology and Cultural History* (BABesch supplement 19): 78-86.

Varone, A. 2008. Per la storia recente, antica e antichissima del sito di Pompei. In P. G. Guzzo and M. P. Guidobaldi (eds) *Nuove ricerche archeologiche nell'area vesuviana (scavi 2003-2006. Atti del Convengo Internazionale, Roma 1-3 febraio 2007*: 349-362. Napoli.

Wallace-Hadrill, A. 2005. Excavation and standing structures in Pompeii Insula I.9. In P. G. Guzzo and M. P. Guidobaldi (eds) *Nuove ricerche archeologiche a Pompei ed Ercolano, Atti del convengo internazionale, Roma 28-30 Novembre 2002*: 101-108. Napoli.

Wallace-Hadrill, A., Guidobaldi M., Camardo. D. and Moesch V. 2008. Le ricerche archeologiche nell'ambito dell'Herculaneum Conservation project. In P. G. Guzzo and M. P. Guidobaldi (eds), *Nuove ricerche archeologiche nell'area vesuviana (scavi 2003-2006)*: 409-424. Roma.

Table: Cesspits of Pompeii

Name of house	Region and insula number	Literature
Casa del Citarista	I 4, 5.6.25.28	Mygind 1921: 317
House of Amarantus	I 9, 11-12	Wallace-Hadrill 2005: 103
House	I 10, 5-6	Ling 1997: 145 note 2
Casa del Menandro	I 10, 4.14-17	own observation
House	I 20, 1-3	Cerulli Irelli 1977: 54-55
Casa di Iulia Felix	II 4	Parslow 2000: 108
Casa degli Epigrammi Greci	V 1, 18.11-12	Staub Gierow 2008: 99-101
Casa delle Nozze d'Argento	V 2, i.e.25	Mygind 1921: 316
Casa di M. Lucretius Fronto	V 4, a.11	Wynia in Peters 1993: 6
House	V 6, 4	own observation
Shop	VI 1, 2.3.4	Hobson 2009: 63-65
Casa delle Vestali (two cesspits)	VI 1, 6-8, 24-26	Hobson 2009: 62-64
Bar	VI 1, 17	Jones and Robinson 2005: 276
House	VI 2, 27	Hobson 2009: 90-91
Casa della Fontana Piccola	VI 8, 23-24	Fröhlich 1996: 70-71 and Abb. 459-460
Casa dei Dioscuri	VI 9, 6-9	Mygind 1921: 317
House	VI 10, 14	Cozzi 1900: 589
Casa del Principe di Napoli (two cesspits)	VI 15, 7.8	Strocka 1984: 18, 21 Abb. 72-74
Casa degli Amorini Dorati	VI 16, 7.38	Seiler 1992: 52-53
House underneath Stabian baths	VII 1, 8.14-17.48.50-51	Eschebach 1979: 52 Taf. 25d
House	VII 2, 18.19	Mygind 1921: 52
House	VII 4, 23-25	Cozzi 1900: 590
Casa degli Capitelli Colorati	VII 4, 31-33, 50-51	Descoedres and Sear 1987: 12, fig. 1, 17 and Sear 2006: 181-184
Casa della Caccia Antica	VII 4, 48.43	Mygind 1921: 317
Casa del Granduca di Toscana	VII 4, 56	Sear 2004: 162-164
Forum	VII 7	Arthur 1993: 193-194
Temple of Apollo	VII 7, 32-33.34-35	Arthur 1993: 193-194

Name of house	Region and insula number	Literature
House	VII 7, 10.13	Cozzi 1900: 587-590
Casa delle Nozze di Ercole ed Ebe	VII 9, 47-48.51.65	Ciarald and Richardson 1999: 75
House	VII 9, 60.63	own observation
House	VII 10, 5.8.13	own observation, belongs to toilet, discovered by Amorso 2005: 52
House	VII 12, 11	Cozzi 1900: 590
Casa di Ganimede	VII 13, 3-4.16-18	Eschebach 1982: 278-280 Abb. 26
Shop	VII 13, 24	Eschebach 1982: 252, 300
House	VII 15, 9.10	own observation
House	VIII 4, 1.53	Dickmann personal communication
Shop	VIII 4, 40-40a	Dickmann and Pirson 2002: 288-289, 303-304
Shop	VIII 4, 44	Dickmann and Pirson 2002: 289-290
Shop	VIII 4, 45 (room 4)	Dickmann personal communication
House	VIII 7, 6	Ellis and Devore 2008: 17-18
Casa di Marcus Lucretius	IX 3, 5.24	Mygind 1921: 317-18
House	IX 6, a.1	own observation
3 cesspits	East of IX, 11	Berg 2008: 364-365
Shop	IX 12, 7	Varone 2008: 354-355; Berg 2008: 366-368, fig. 5
Casa dei Casti Amanti	IX 12, 9	Varone 2008: 354-356, fig. 8

The latrine at the Roman fort on the Antonine Wall at Bearsden

David J Breeze

The fort

The existence of a Roman fort on the Antonine Wall at Bearsden (Figure 1) was recognised in the 18th century. It was planned by William Roy in 1755. During the 19th century the fort was progressively submerged under the housing of the expanding Glasgow conurbation. About 1970, the four Victorian villas occupying the northern half of the fort were acquired for a new development. In 1973, I undertook an exploratory excavation on behalf of the predecessor of Historic Scotland in order to determine what, if anything, survived of the fort. The defences were located, together with internal buildings. One such building was the bath-house. Following preliminary excavation, this was covered over, to be reopened in 1979 prior to completion of excavation and consolidation as a publicly owned monument (Breeze 1984).

Excavations were undertaken within the area to be developed and within the grounds of a house south of the road so that a plan of the fort and its annexe could be interpolated (Breeze 2016). During the course of examination of the eastern defences, it was noted that the fill of the outer ditch (the inner ditch was inaccessible) smelt. It was not until a subsequent season that the reason was discovered through the location of the latrine. The drain from the latrine fed into the fort/annexe ditches.

The excavations were completed in 1982. It was determined that the fort was occupied during the middle years of the second century. The building of the Antonine Wall started in 142, but the fort at Bearsden was secondary and may not have been constructed until some years later. It is believed that the Antonine Wall was abandoned in the 160s.

The latrine

The latrine lay to the south of the south-east corner of the bath-house within the annexe (Figure 2, see Breeze 2016: 70-2; 327-30). Various drains and a path led to the latrine from the bath-house. The gravel path was traced from the entrance to the bath-house round the exterior of the south side of the building to the south-east corner of the hot bath where it turned south to the entrance of the latrine. Here, it overlay the drain from the stoke-hole.

The latrine (Figure 3) measured 5 m by about 4.4 m: the south wall had been destroyed so a slight element of doubt surrounds the north-south measurement. The west face of the east rampart of the annexe formed the east side of the building, but the north wall extended 0.2-0.3 m over the rampart kerb. The latrine wall did not lie directly on the kerb stone, but on a layer of yellow/brown clay mixed with sandstone fragments 0.20 m thick which overlapped the edge of the kerb stone and

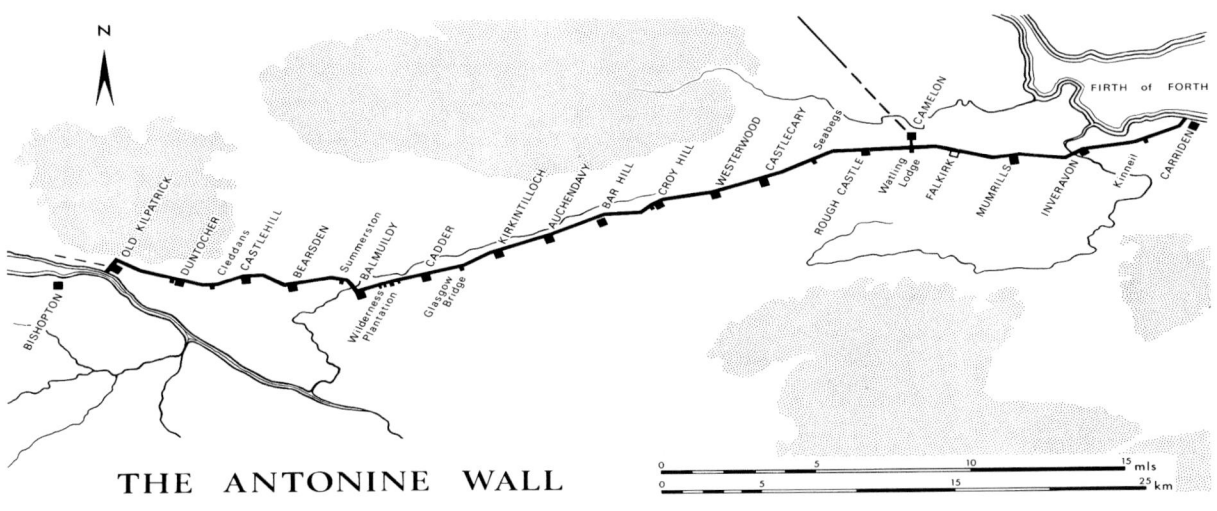

Figure 1. Plan of the Antonine Wall (after Breeze 2016, Illustration 1.1)

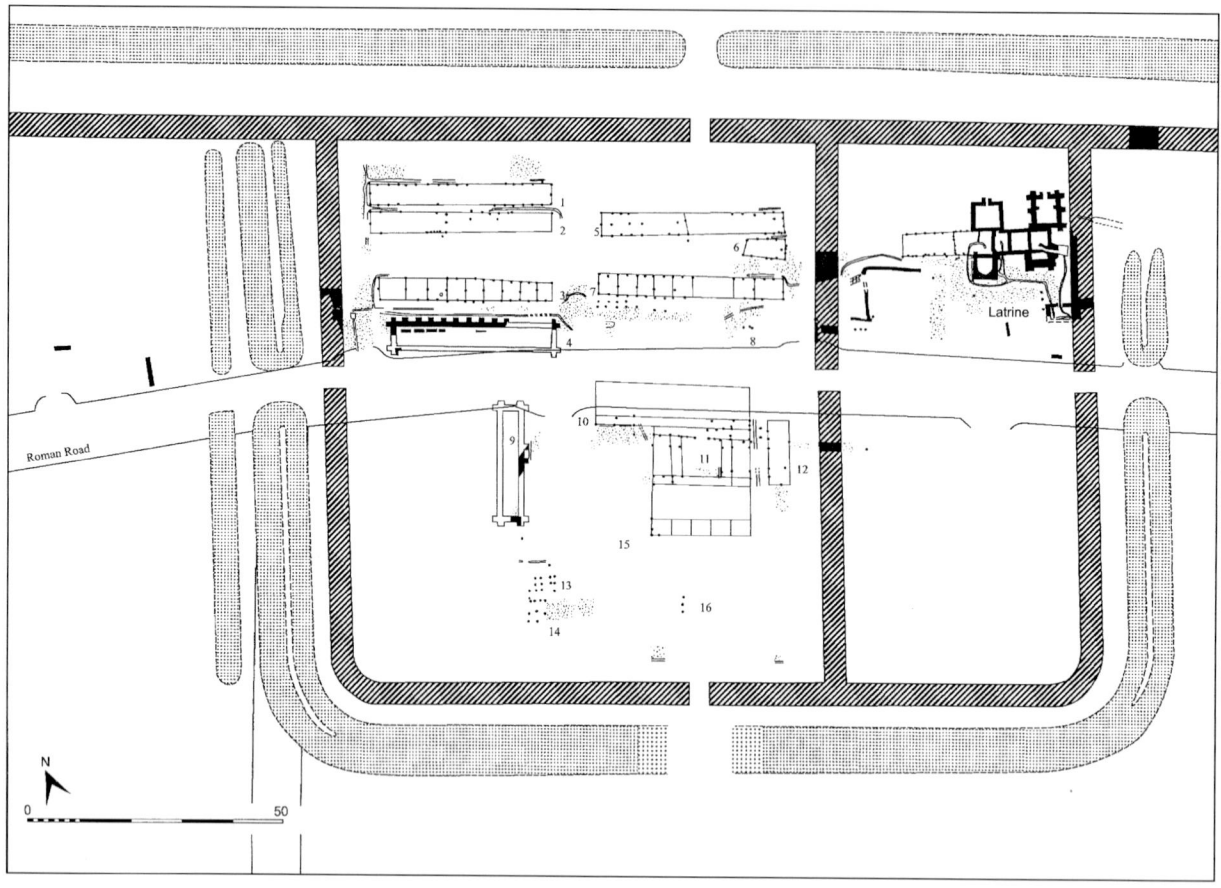

Figure 2. Plan of the fort and annexe at Bearsden; the latrine lies immediately inside the east rampart of the annex, south of the bath-house (after Breeze 2016, Illustration 3.2.1)

which in turn lay on top of the underlying grey clay into which the rampart kerb had been cut.

The walls of the latrine varied from 0.6 to 0.7 m thick. They were formed of roughly coursed rubble. The entrance into the building lay in the north wall. It was placed 1.5 m from the inside north-west corner and was 1 m wide. The threshold was formed of three flags (Figure 4).

The entrance led into a paved area measuring about 3 by 1.5 m. Between this paving and the west and south walls of the building lay two channels. Immediately inside the walls of the building lay the main channel. That on the west side was unusually shallow, only about 100 mm deep and a maximum of 400 mm wide, with a flagged floor. Although the south channel had been damaged by the modern sewer pipe, it could be determined that it also was shallow but was not flagged: the bottom was cut into the natural clay. This channel passed through the rampart in a well-constructed drain with a flagged bottom, sides two courses high supporting cap-stones.

Inside the sewage channel lay an open channel cut into a series of long stones. This was fed by the channel into which the five drains from the bath-house de-bouched. This channel passed through the north wall of the latrine. Beneath the end of the channel, on the inside of the north wall of the latrine, a tile had been positioned at an angle so as to throw water into the channel below. It would make sense if the tile had been supported by a device so that water could have been directed into the main sewage channel or the open channel.

The floor of the latrine lay on orange/brown clay, in which were embedded, on the south side, masons' chips. The drain leading south from the stoke-hole was cut through this clay. It entered the latrine under the door, crossed the building diagonally and sinuously to exit into the south-east corner of the main channel. This narrow drain, 120 mm deep, was roughly constructed mainly with water-worn stones or rough boulders. There were no cap-stones: the flags above appeared to have been bedded in clay laid over the top courses of the drain.

A thin layer of rubble covered the building, lying thicker to the west end than beside the rampart. This material contained charcoal up to 50mm in diameter, roundwood of alder, hazel and willow probably the

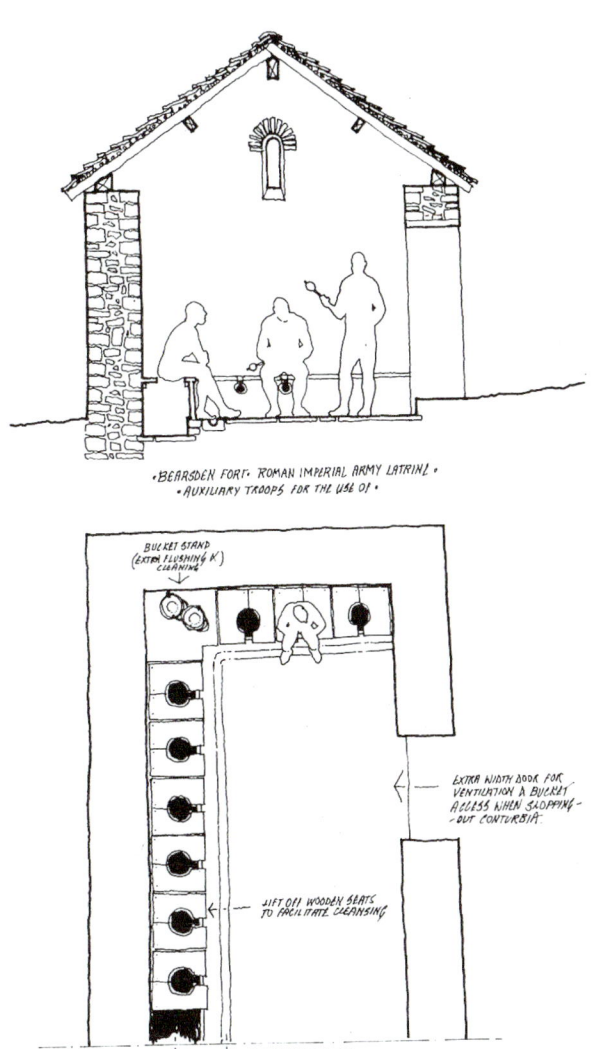

Figure 3. Plan and section of the latrine. Drawn by Michael J. Moore (after Breeze 2016, Illustration 21.10)

Figure 4. The latrine under excavation looking west (after Breeze 2016, Illustration 3.3.58)

remains of wattles, and small fragments of mainly clay-covered burnt rushes, all presumably from the superstructure. The main channel contained gravel, the normal fill at the bottom of the drains. Here also were found numerous rush seeds, possibly from rush thatching on the roof (Dickson and Dickson 2016: 267).

The sewage

The outer ditch of the annexe was sectioned at one point and a column sample taken of the contents. On analysis by Camilla Dickson at the Department of Botany in the University of Glasgow it was realised that this was sewage (Dickson and Dickson 2016: 223-8). Subsequent work was undertaken by Dr Brian Knights (Knights *et al* 1983).

Within the sewage, fragments of hulled wheats, probably from both emmer and spelt wheat, were identified, together with bran fragments from either wheat or rye: the bran formed about half the organic part of the ditch infilling. Rye and oats also appear, but may have been merely weeds in a wheat crop. In addition, the sewage contained barley grain fragments which had been ground with the wheat and also fragments which had been processed in a similar manner to pearl barley. The bulk of the grain would appear to be emmer and spelt with a little barley.

It seems likely that the wheat was imported to Bearsden rather than grown locally, and this receives support from the presence of grain beetles in the sewage (Locke 2016: 289-97). The grain beetles could have either entered the ditch by the dumping there of contaminated grain, or through the soldiers eating contaminated grain. Isolated fragments of grain beetles were found elsewhere on the site and their presence in the sewage may be thought more likely to be the result of their being eaten rather than that this ditch was coincidentally used for the dumping of contaminated grain. The beetles may have been eaten within the pearl barley used to thicken soup (Breeze 2016: 369-70).

Other foods were found in the sewage. These include lentil, horse bean, linseed, fig, dill, coriander and opium poppy (Dickson and Dickson 2016: 231-2). Lentil may

Figure 5. The latrine following consolidation and presentation looking north (after Breeze 2016, Illustration 22.3)

have been imported to Bearsden from southern Britain and fig, dill, coriander and opium poppy from the continent. Wild plants eaten at Bearsden include wild celery, wild turnip, wild or cultivated radish, common mallow, bilberry, wild strawberry, blackberry, raspberry, hazel nuts and purging flax.

The biochemical analysis undertaken by Knights hinted that the soldiers had a mainly vegetarian diet (Knights *et al* 1983). This work was particularly important because, while it is known that Roman soldiers ate meat (Davies 1971: 126), the balance of meat within the diet was unknown. The work of Knights demonstrates that the vegetarian part of the diet was more important that meat. Unfortunately at Bearsden the ph level of the soil (5.6) had resulted in the destruction of all bone.

Finally, it may be noted that the soldiers had worms. Examples of both *trichuris trichiura* (whipworm) and *ascaris* were present in the sewage. A single example of a human flea was also recovered from the sewage (Jones and Maytom 2016).

Comment

It may be presumed that the main sewage channel would have been covered by the wooden seating and a deeper channel may therefore have been thought unnecessary. There was space for nine seats. The channels in the latrine were fed by water from the bath-house. The sewage channel passed through the rampart and emptied its contents into the annexe ditch. The discovery of fragments of mosses in the outer annexe ditch suggests that this material may have been used for personal cleanliness, being dipped into the open channel running round the interior of the latrine (Figure 5).

Bibliography

Breeze, D. J. 1984. The Roman fort on the Antonine Wall at Bearsden, In Breeze, D. J. (ed), *Studies in Scottish Antiquity*: 32-68. Edinburgh.

Breeze, D. J. 2016. *Bearsden: A Roman Fort on the Antonine Wall*. Edinburgh.

Davies, R. W. 1971. The Roman Military Diet, *Britannia* 2: 122-142.

Dickson, C. A. and Dickson, J. H. 2016. Plant remains, in Breeze 2016: 223-80.

Jones, A. K. G. and Maytom, J. 2016. Parasitological investigations of the east annexe ditch, in Breeze 2016: 301-3.

Knights, B. A., Dickson, C. A., Dickson, J. H. and Breeze, D. J. 1983. Evidence concerning the Roman Military Diet at Bearsden, Scotland, in the 2nd Century AD. *Journal Archaeological Science* 10: 139-152.

Locke, J. 2016. Insect remains, in Breeze 2016: 289-99.

Flushed with success – a Roman flushing installation in the latrines of the Great Bathhouse of the Colonia Ulpia Traiana near Xanten (D)*

Norbert Zieling

The public bathhouse complex on the western *insula* no. 10 of Colonia Ulpia Traiana had originally had two latrines (Figure 1). One of them was situated at the NE side of the *insula*, outside of the bathhouse complex proper and was most likely only accessible from the street, but not from the bathhouse itself (Figure 2) (Zieling 1999; Zieling 2008). The visitors of the bathhouse could avail themselves of a latrine in the southern corner of the *insula*, which was accessible both from the bath's *palaestra* and via two rooms appended to the *basilica thermarum*, the larger of which is interpreted as an *apodyterium*. The rectangular toilet room had a size of around 124 m². A channel ran parallel to all four walls, supplied with water from the SW cold-water pool of the bathhouse via an opening on the NE side. A striking feature was noticed at the excavation in 1963 below the

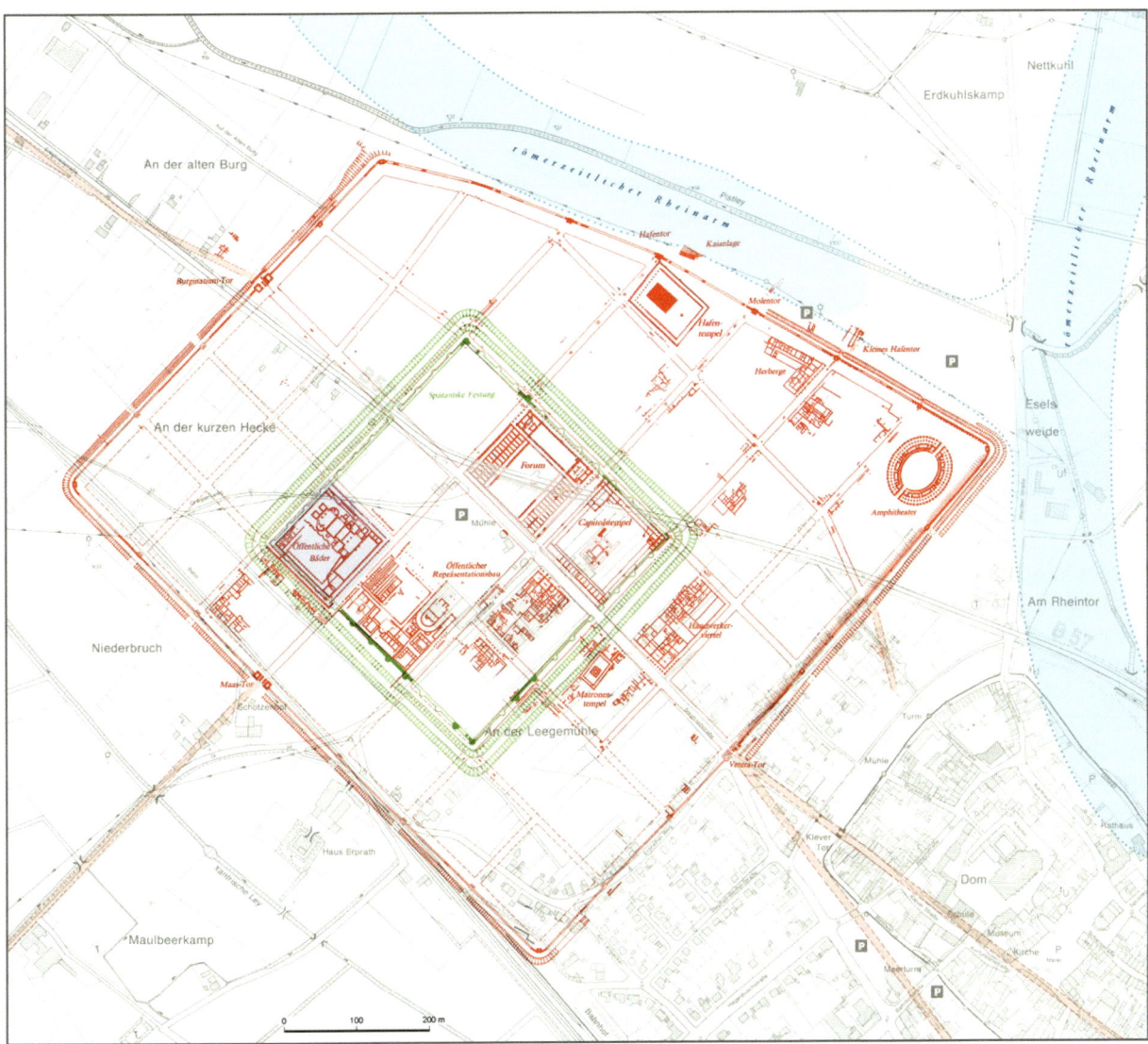

Figure 1. Xanten, Colonia Ulpia Traiana. Plan of the features, with the bathhouse complex in the western part of the city indicated. Graphics Horst Stelter, LVR-Archäologischer Park Xanten / LVR-RömerMuseum

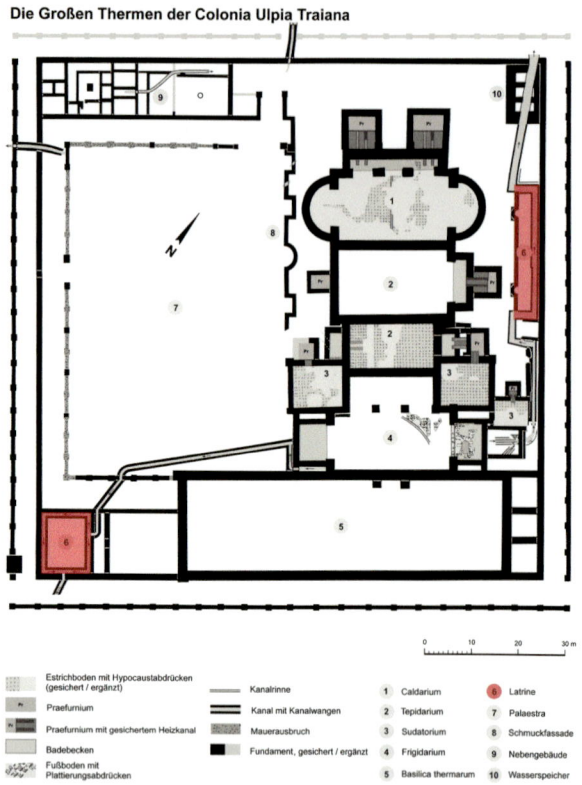

Figure 2. Xanten, Colonia Ulpia Traiana. Ground plan of the bathhouse complex with latrines indicated. Graphics Horst Stelter, LVR-Archäologischer Park Xanten / LVR-RömerMuseum

Figure 3. Xanten, Colonia Ulpia Traiana. Niche in channel bottom, excavation 1989/90. LVR-Archäologischer Park Xanten / LVR-RömerMuseum.

opening: the ring-channel of the bathhouse latrine had a niche of 5,30 m long right at the water supply opening, which necessitated an interpretation (Figure 3).

The first excavation of the latrine was in 1879, when members of the Niederrheinischer Altertumsverein Xanten excavated parts of the bathhouse complex, including the latrine (aus'm Weerth 1880: 69-70). The excavation plan published by Alfons de Ball in 1882 proves that the whole of the latrine had been excavated (de Ball 1882: Taf. IV) The plan also reveals a large rectangular foundation inside the ring channel at the SE side of the room. According to the excavator, this foundation reached almost to the ground level. Presumably, this was the substructure for a fresh water reservoir, conducting the water used for cleaning oneself into the little channel which must have ran on the floor at the feet of the latrine users. Unfortunately, none of these features could be rediscovered during the excavations of the 20th century.

In contrast to this, the ca. 0,60 m wide sewage channel running along the walls was found completely preserved during the excavations of 1989/1990. Levelling on the bottom surface of the channel has established that the water was lead through the whole ring channel along all four walls with a minimal gradient only. The ring channel exited the room on the SE side of the latrine, from whence the sewage was led via a channel turning SE into the sewer of the *decumanus maximus*.

An attempt to correlate the different heights of several contexts in the bathhouse complex and in the latrine itself resulted in an approximate height for the ring channel of 1,60-1,70 m, if the seating height is assumed to have been 0,50 m.

The aforementioned niche is 5,30 m long and sits beneath the water feed directly in front of the NE wall of the latrine. With a width of 0,20 m, it takes up slightly more than a third of the width of the ring channel. Because of the narrowness of this recess, the filling of mixed humus could only be excavated to a depth of around 0,40 cm. The narrow ends of the niche were approximately half round. The walls of the toilet room were plastered to the height of the ring channel with plaster mixed with crushed brick, but this plaster was missing on the wall above the niche. Because the edges of the plaster exactly correspond with the outer corners of the niche, it is obvious that the wall above the niche had never been plastered. It seems logical that something in the nature of an installation or some facing must have covered the whole length of the wall (5,30 m) in front of the niche. This is supported by the documentation of the 1989/1990 excavations, which in detail record the plaster remains noted at that time.

The location directly at the water feed of the latrine makes it plausible that the installation that sat here was an overflow container, which was embedded in the niche on the bottom. If this container was from metal and had been used as means to collect the overflow water from the cold-water pool of the *frigidarium* some 45 m distant, we need to consider how this worked technically. In order to do this, the feed conditions will have to be regarded first.

Examination of the wastewater drains at and on the *frigidarium* clearly demonstrates that the two main drains which took up the wastewater from the cold-water pool also took up the water from the roof of the two covered passages between the *basilica thermarum* and the *frigidarium*. On rainy days, the water gathered this way must have amounted to a considerable quantity. In order for the latrine flushing to work faultlessly even during dry spells, the ring channel had to be flushed continuously with at least a small amount of water from the pool. This again required the cold-water pool to continuously receive fresh water from the fresh water supply pipe. It seems likely that the cold-water pool was fitted with an overflow, which led the water via a plastered of maybe even a grilled opening in the SE wall into the channel.

Another possibility could have been an open channel running along the rim in the height of the surface of the water, as in modern pools, where this channel usually is fashioned as a handhold for the bathers. As the upper part of the pool has not been preserved, it is alas impossible to prove either theory.

An important component in judging the technical features of the latrine is the evidence of the water levels. The lower edge of the drainage opening in the cold-water pool is at a level of 21,96 m above sea level. If we assume an average water depth of c. 1 m in the pool, which is known from several Roman bath buildings, the water level is at c. 23 m above NN (= Normalnull, a vertical measurement referring to the height above mean seal level). The water leaving the pool ran through the drain in the *palaestra* with a relatively even gradient to a level of 21,40 m above NN at the opening into the ring channel. The difference (water level 23 m above NN – upper edge channel bottom in the latrine 21,40 m above NN) results in a water height of c. 1,60 m.

If we assume that this water height of 1,60 m was used to the maximum, the overflow container should have had the same height. If we further assume the width

Table

Capacity Overflow Container	Container high and narrow	Container high and wide	Container low and narrow
Water level cold-water basin	~ 23,00 m above NN	~ 23,00 m above NN	~ 23,00 m aboven NN
Bottom level latrine	21,40 m above NN	21,40 m above NN	21,40 m above NN
Length of channel / niche	~ 5,30 m	~ 5,30 m	~ 5,30 m
Height Water column	~ 1,60 m	~ 1,60 m	~ 0,70 m
Width of channel / niche	~ 0,20 m	~ 0,60 m	~ 0,20 m
Calculated base area	~ 1,06 m²	~ 3,18 m²	~ 1,06 m²
Calculated volume of container	**~ 1,7 m³**	**~ 5,1 m³**	**~ 0,75 m³**

of the overflow container (corresponding to the width of the niche) was 0,20 m, the height 1,60 m and the length 5,30 m, the filling volume of the container was around 1,7 m³ water (Table). A second variant would be a vessel of 1,60 m, which would use the full 0,60 m width of the channel. This calculation results in a capacity of more than 5,0 m³. The latter variant is the less probable one, as this would mean that the channel would have been useless at this point and that it would have been impossible to install seats above the vessel, as otherwise faeces would have gotten into the vessel.

Probably the best indication of the real height of the overflow container can be gained from the excavation report of 1989/1990. Slight remains of the crushed-brick plaster on the NE wall of the latrine demonstrate that the wall had been plastered above the niche, the lower edge of this plaster being at 22,08 m above NN. The resulting height of the overflow container between channel bottom and lower edge of the plaster is just almost 0,70 m. If we re-calculate the amount of water stored in the container, we can conclude that if the overflow container had a height of 0,70m, it could have had a capacity of c. 0,75 m³ water, which is still 750 l water. If we consider that a modern toilet flush container has a capacity of 6-9 l water, it seems conceivable that the amount of water we calculated as capacity of the overflow container would have been enough to flush the whole latrine with up to 70 seats on opening the hatch.

As the seats were not preserved, the number of seats was determined by measuring the distance between the seats in comparable installations elsewhere, whose seats had been preserved. The number of 70 seats is based on the assumption that all four walls of the latrine had the maximum possible number of seats.

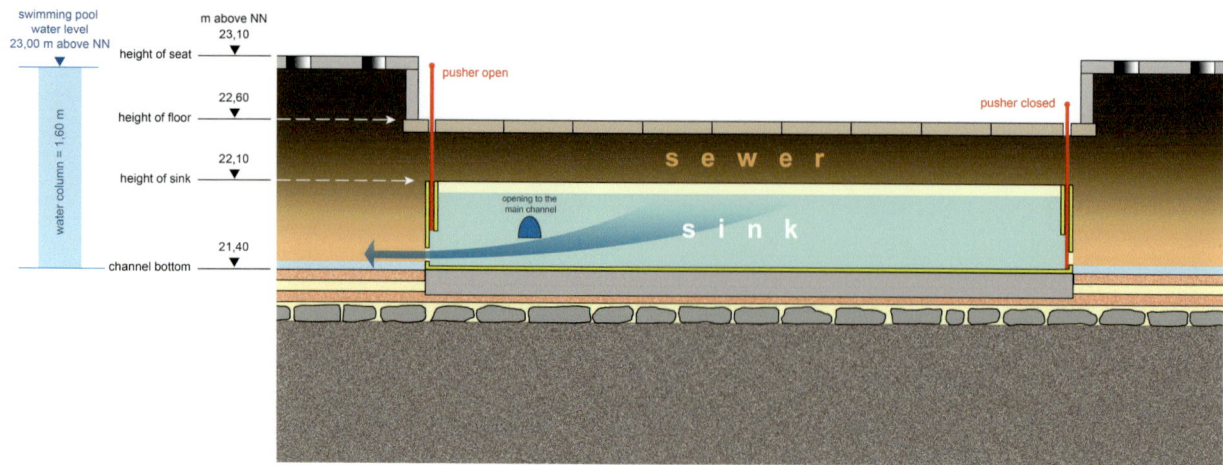

Figure 4. Xanten, Colonia Ulpia Traiana. Tentative reconstruction of the overflow container, front view. Graphics Horst Stelter, LVR-Archäologischer Park Xanten / LVR-RömerMuseum

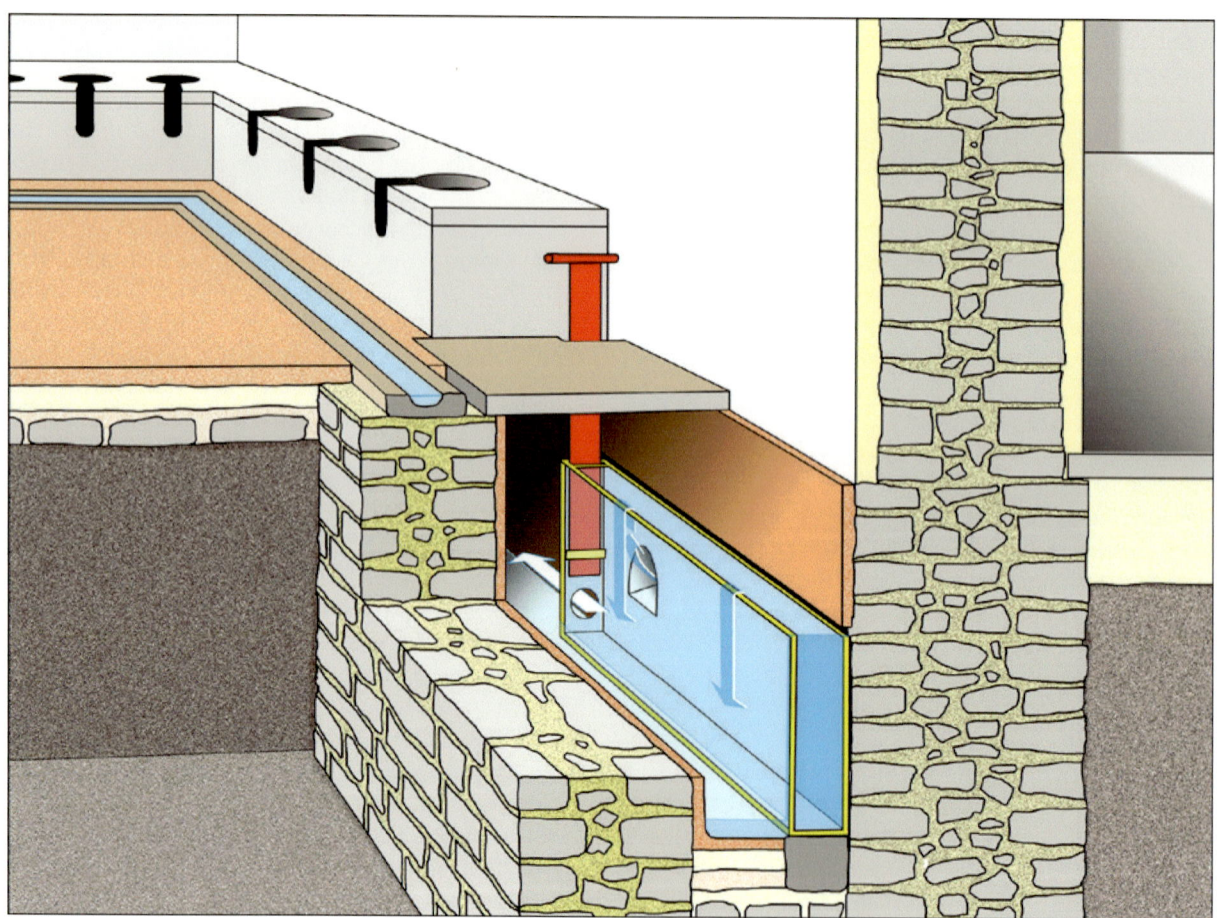

Figure 5. Xanten, Colonia Ulpia Traiana. Tentative reconstruction of the overflow container, section. Graphics Horst Stelter, LVR-Archäologischer Park Xanten / LVR-RömerMuseum

How such a hatch might have looked is alas unknown. The following illustrations are thus only to be seen as attempts at a reconstruction of the situation (Figures 4, 5). The problem of possible leaks at the hatches pictured in red would have been negligible as we are only talking about wastewater. The reconstruction drawings presented here assume that the mechanism to operate the flushing was situated in the latrine chamber itself, among the customers. It seems entirely conceivable – even likely – that the mechanics had been transferred by some lever arms through an opening in the wall to the back of the wall. The passage connected to the latrine at the NE wall seems predestined for such a function. From here, the latrine could have been flushed regularly almost unnoticed by the guests.

(translated by Stefanie Hoss)

Bibliography

Ball, A. de 1882. Bericht über die Ausgrabungen auf der alten Burg zu Xanten bis Mitte November des Jahres 1881. *Bonner Jahrbücher* 74: 76ff.

Weerth, E. aus'm 1880. Vorläufiger Bericht über die neuen Ausgrabungen bei Xanten. *Bonner Jahrbücher* 69: 68ff.

Zieling, N. 1999. *Die Großen Thermen der Colonia Ulpia Traiana. Die öffentliche Badeanlage der römischen Stadt bei Xanten* (Führer und Schriften des Archäologischen Parks Xanten 19). Köln.

Zieling, N. 2008. Die Thermen. In M. Müller, H.-J. Schalles and N. Zieling (eds), *Colonia Ulpia Traiana. Xanten und sein Umland in römischer Zeit. Geschichte der Stadt Xanten* (Xantener Berichte Sonderband 1). Mainz.

Zieling, N. 2009. Eine römische Toilettenspülung in den Großen Thermen der Colonia Ulpia Traiana. *Xantener Berichte* 15: 313-320. Mainz.

The latrines of Roman Aachen

Andreas Schaub

Introduction

In the Roman period, Aachen was probably called *Aquae Granni*. The combination of *Aquae* – water and the name of the Celtic deity *Grannus,* a god of healing and mineral springs aptly describes the foundation cause: The local hot sulphurous springs were the reason the Romans founded a settlement at this place some time around the birth of Christ.

One of the first construction activities in Aachen seems to have been the erection of a thermal bathhouse at the Quirinus spring (Sage 1982: 86 f; Hugot 1982: 131 f. For a new and comprehensive re-evaluation of Roman Aachen see Schaub 2012). Shortly thereafter, around the mid-1st century AD, the thermal complex of the so-called 'Büchelthermen' followed, which were fed by the Kaiser spring (Cüppers 1982: 47-48). Only after the oldest thermal bathhouse had been abandoned in the 2nd century AD, a second new large thermal complex was build, the so-called 'Münsterthermen'.

The settlement activities in *Aquae Granni* were not limited to thermal baths, even though these installations seem to have been the strongest economic factor of the town. The settlement area amounted to about 16 to 20 ha. With an orthogonal street layout and evidence for large stone buildings, Roman Aachen was a town with quite an urban 'look'. A large amount of evidence for trade and crafts paint a picture of a prosperous town in the centre of the province Germania Inferior. Finds indicate a continuous settlement through to the Early Middle Ages (Schaub 2008).

At the moment, there are no recent excavations or new research on the latrines of Roman Aachen. They nevertheless existed, and at least two were published in the past. These and a possible third latrine shall be briefly presented here (Figure 1).

Latrine 1

During some ground construction work in the years 1958 and 1962, Leo Hugot was able to discover part of a large latrine installation at the 'Büchel'. I would like to shortly relate only the more important details from Hugot's extensive description of the building here (the following after: Hugot 1961: 85-89; Hugot 1963: 196; Cüppers 1982: 53-55): He discovered a building oriented EW and 19,50 m long, whose width could be documented to at least c. 4,60 m (Figure 2). The building lies above a generously dimensioned wastewater drain with a width of 0,60 m and a height of 0,95 m. The gradient from west to east amounts to around 2% (Hugot 1959: 87). If one follows Hugot and assumes the main drain was the centreline, the original width of the building might have been almost 10 m. At the western end of the latrine, a side channel lined in *opus signum* branches off from the main drain. This flushing channel was responsible for the removal of the faeces and rejoined the main drain at the eastern end of the building, some 0,14 m deeper than at the point of branching off. The outer wall of the latrine doubled as one of the channel sides and with almost 0,80 m is very wide. The inner second side of the channel on the other hand just measures 0,45 m. The *opus signinum* of the flushing channel was continued above the top of the inner wall and ended at a small open drain formed by large blocks of Aachen bluestone (a blueish limestone found in Aachen, dissimilar to English bluestone, a kind of dolerite). Directly in front of this bluestone drain, another trough-like drain had been inserted into the floor finish. The outer wall reached 1,13 m above the upper edge of the bluestone drain and had a recess of 0,12 m width at ca. 0,40 m above the drain. Like the inner wall, the upper edge of the outer wall was plastered with a plaster containing crushed brick.

The interpretation as latrine is certain, however Hugot's proposal to see the building as open or covered with only a light wooden construction is unlikely. The discovery of a sort of floor finish of the outer wall seems to point at the smoothed-out finish of a base wall, on which the foundation beams of a half-timbered construction rested. The recess in the outer wall was used as support for the seats (Figure 3). One may assume a height of c. 0,45 m for the seats.

The size and location of the building point to the interpretation as a public latrine. Even though no actual remains of a street could be detected in front or next to the building, the public use is indicated by existence of the main drain in the latrine (Figure 2). The street running EW along the latrine may have had a width of 3,20 m at that point, if one counts a wall running parallel to the latrine as the front of a building on the opposite, northern side. Another possibility is that the road was wider to the east and west of the latrine, narrowing in the area of the latrine.

The distance between latrine and the thermal complex 'Büchelthermen' is more than 20 m. No direct

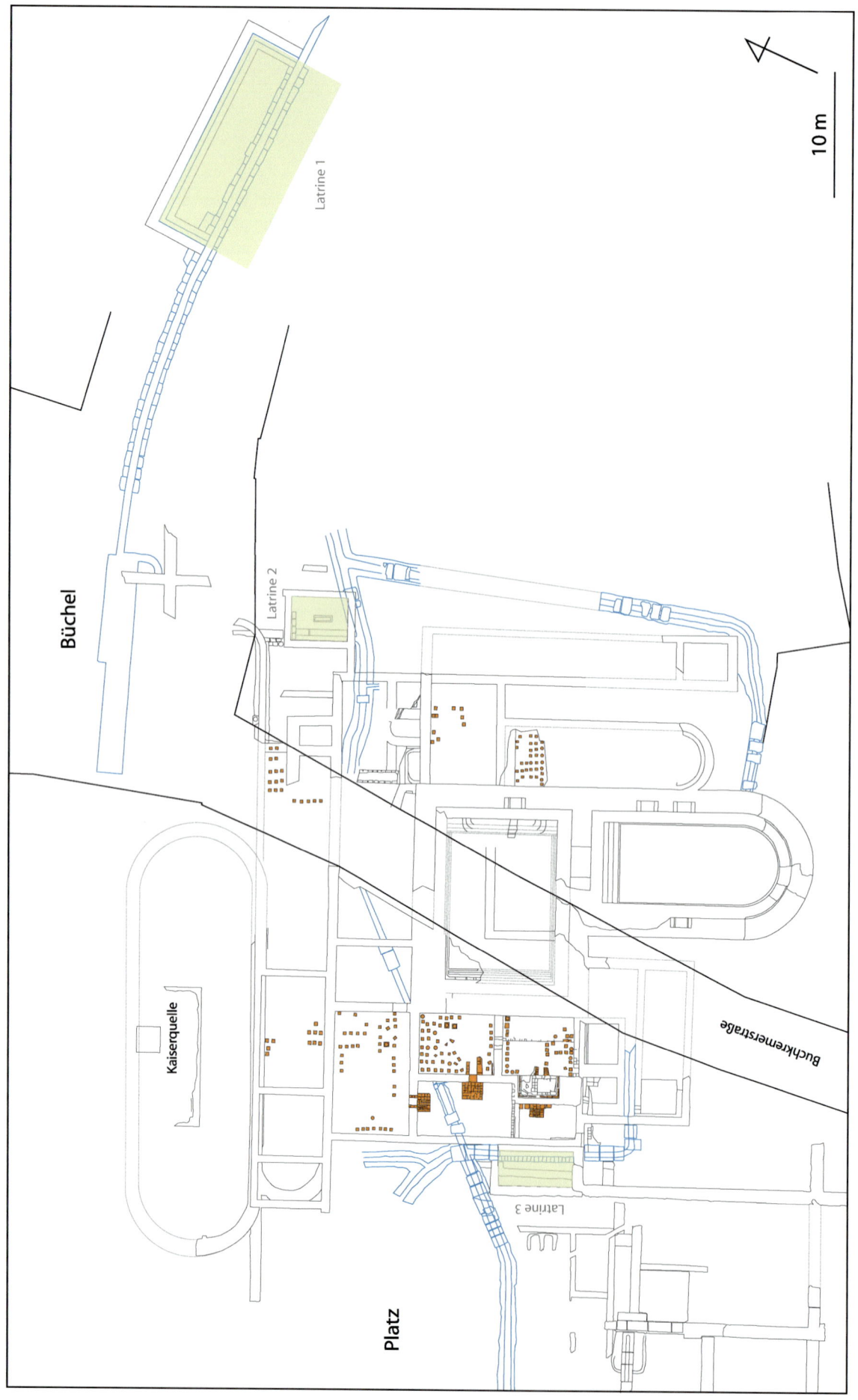

Figure 1. Schematic and reconstructed plan of the Roman stone buildings at the 'Büchel' with highlighting of the latrines (light green grid). M 1:400 (after Hugot 1982, Tafel 57, supplemented by author).

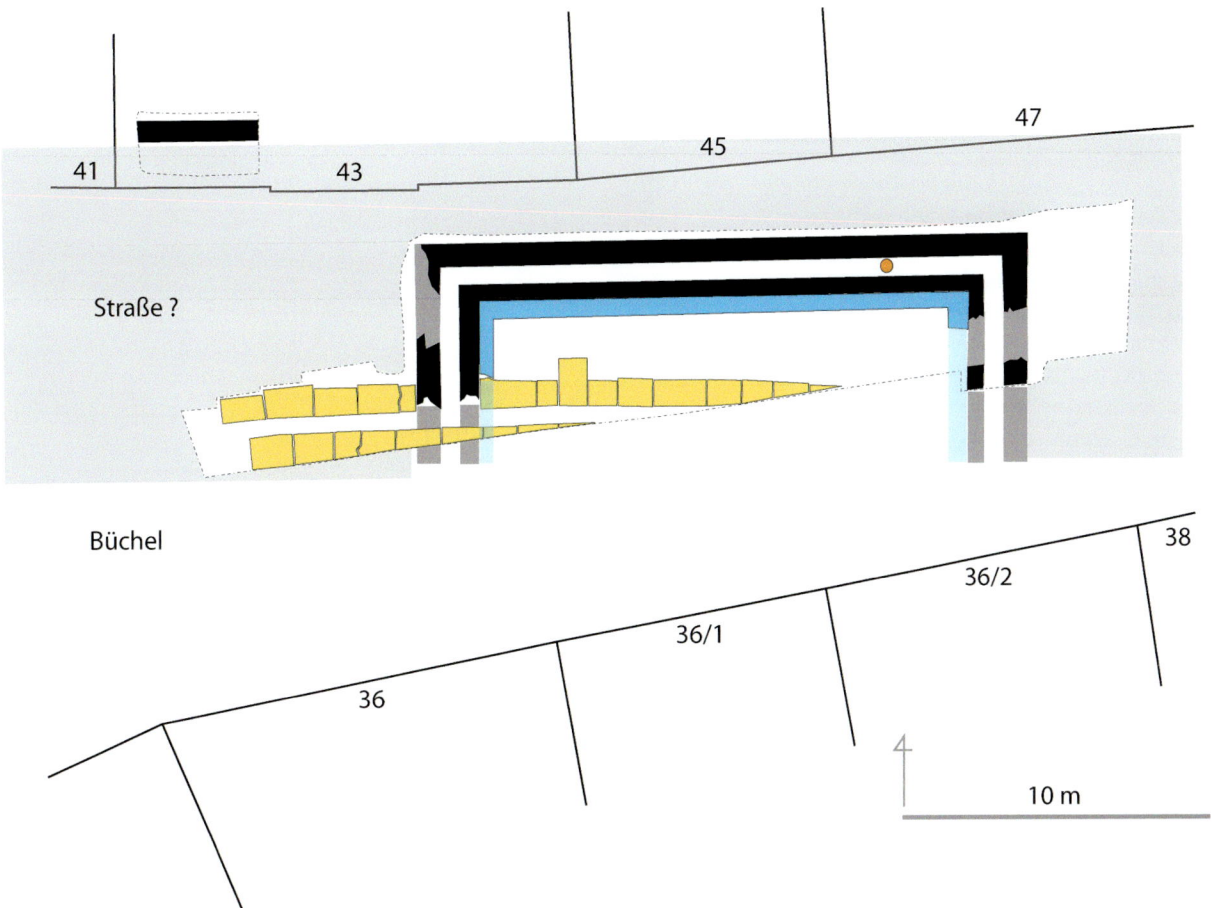

Figure 2. Public latrine at 'Büchel'. M 1:200 (after Hugot 1963: 195, Abb. 5, supplemented by author).

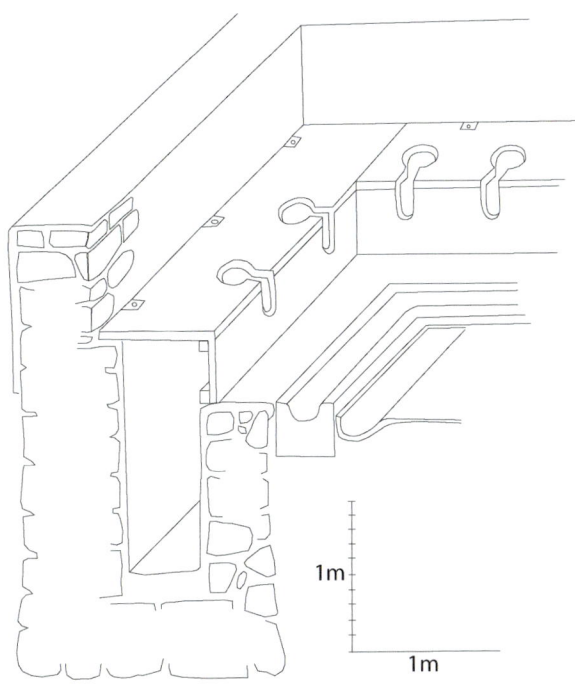

Figure 3. Tentative reconstruction of the public latrine at 'Büchel'. M 1:50 (after Hugot 1961: 89, fig. 80)

constructional connection exists. Nevertheless it cannot be excluded that the latrine was the northern terminal of a larger building complex yet unknown.

But is seems more likely that the location of the latrine is connected to the extensive existing drain system in the area, with all drains ultimately leading into the main drain (employed by the latrine), which ensured a strong permanent water current for flushing.

The public latrine is situated on the eastern edge of the settlement. No remains of buildings have been discovered to the east of the latrine to date. Nevertheless, the sheer size of the installation indicates that the area was much frequented, a point borne out by its location directly to the north of the thermal complex and to the NE of the central porticoed square of *Aquae Granni*. A public latrine of these dimensions is a clear sign of the urban character of Roman Aachen.

The only find indicating a date is the small complete jug dating from the second half of the 2nd to the first half of the 3rd century AD, which was found in the flushing channel (Hugot 1963: 196, Abb. 6). Its find location is marked by a red dot on Figure 2. The jug must have fallen into the flushing channel through one of the seat openings. It was probably used for cleaning oneself after the use of the latrine.

Latrine 2

The second installation that may be confidently described as a Roman latrine was also excavated and published by Leo Hugot (1982: 145-147 and 156). Contrary to the first latrine, latrine 2 was integrated into the complex of the thermal complex 'Büchelthermen' (Figure 1). The room was located in the northern corner of the thermal complex. Unclear remains of walls and floors to the NW and NE indicate, that the latrine was not necessarily the last room of the building in this wing. As the entrance was from the north, the latrine could not be entered directly from the bath. The peripheral location indicates that the latrine could also be used by passers-by not staying in the thermal complex.

The eastern corner of the room rests directly on the stone covers of a main drain running SE with a gradient to NE (Figure 4).

The room measures 5,60 x 4,20 m, with the outer walls being c. 0,50 m wide. The flushing channel with a bottom of *tegulae* runs along the SW and NW-side. It is 0,45 m wide on the inside and the maximum depth at the entrance with the main drain is more than 0,60 m. While the drainage is thus apparent, the main water supply is still unknown. It may have been from the north, where the ancient situation has been obscured by later superimpositions. The entrance of a small side channel made by two *imbrices* has been documented in the NE-side of the channel.

The floor of the latrine is made up of lime mortar. A monolithic bluestone basin, 1,60 m long and 0,40 m wide was placed in the centre of the room on the longitudinal axis. It was maximally 0,10 m deep and had an overflow at the narrow side in the SE. It probably was used for the same purpose as the open bluestone drain in latrine 1.

While at least the NW outer wall was decorated in reddish plaster, the inner walls were painted polychromatic. Above the white ground coat, remains of ochre, red and green pint have been preserved in the order they fell from the wall.

Latrine 3

Another context from the thermal complex 'Büchelthermen', excavated in 1956 by Wilhelm Lehmbruck will be presented here as possible third

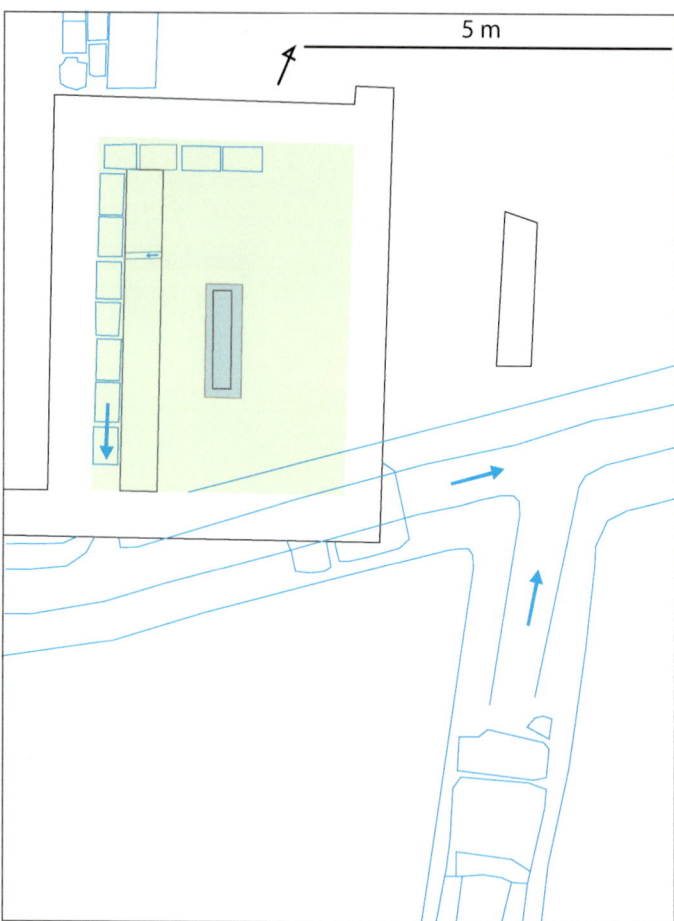

Figure 4. Latrine 2. Detail from fig. 1. M 1:100 (after Hugot 1982, pl. 57, supplemented by author).

latrine (Figure 1, Lehmbruck 1958). The room of 7,50 x 4,50 m (Figure 5) is located on the outside of the SW wall of the thermal complex and was described by Lehmbruck (1958: 33, Abb. 1, room 10) in his excavation report as 'annexe' with no further suggestions to its function. Heinz Cüppers (1982: 43) includes it in period 2 of the thermal complex, in which another large drain had been constructed. This drained the thermal basins in the south of the complex, ran along the long NE-side of the presumed latrine and directly to the north of it met the main drain coming from the SE (Lehmbruck 1958: 37, Abb. 13). The interpretation of the room as a latrine put forward here rests on the peripheral location in relation to the whole complex and the fact that a large drain was led through the whole room on the longitudinal axis. Further installations are not documented.

Conclusion

Both latrines known in Roman Aachen to date have been excavated, interpreted and published by Leo Hugot. Another possible latrine has been added in this article, raising the number to three. While latrine 2 and the possible latrine 3 clearly belong to the large thermal building complex of the 'Büchelthermen', latrine 1

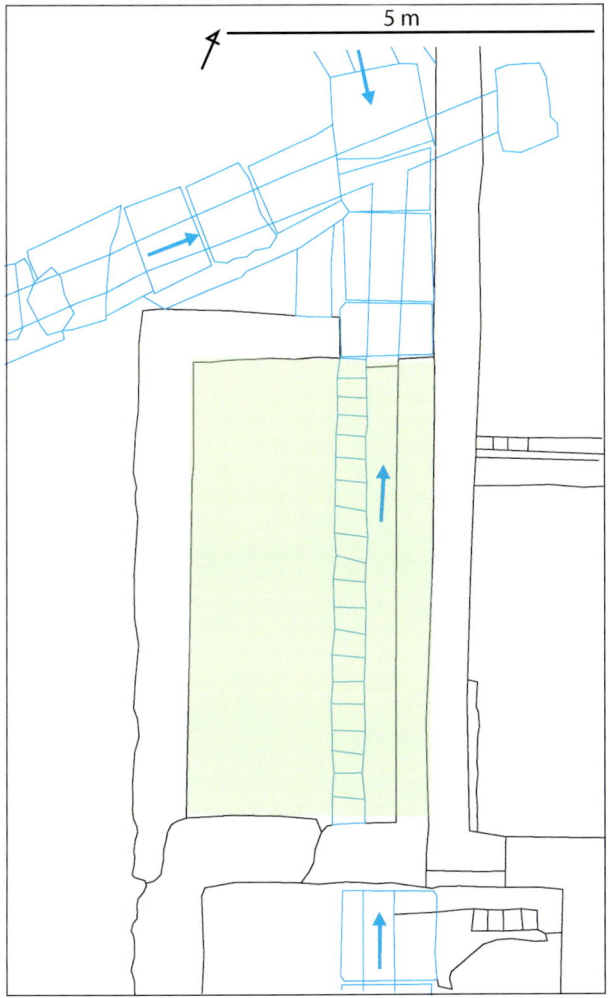

Figure 5. Latrine 3. Detail from fig. 1. M 1:100 (after Hugot 1982, pl. 57, supplemented by author).

Bibliography

Lehmbruck, W. 1958. Ausgrabungen in einer römischen Therme in Bad Aachen. *Aachener Adressbuch* 1957/58: 33-39.

Hugot, L. 1961. Neue Forschungen zur römischen Bücheltherme in Aachen. *Aachener Kunstblätter* 19/20 (1960/61): 85-89.

Hugot, L. 1963. Die römischen Bücheltherme in Aachen. *Bonner Jahrbücher* 163: 188-197.

Cüppers, H. 1982. Beiträge zur Geschichte des römischen Kur- und Badeortes Aachen. In *Aquae Granni. Beiträge zur Archäologie von Aachen.* (Rheinische Ausgrabungen 22): 1-75.

Hugot, L. 1982. Ausgrabungen und Forschungen in Aachen. In *Aquae Granni. Beiträge zur Archäologie von Aachen. Rheinische Ausgrabungen* 22: 115-173.

Sage, W. 1982. Stadtkerngrabungen in Aachen 1962-1964. In *Aquae Granni. Beiträge zur Archäologie von Aachen. Rheinische Ausgrabungen* 22: 77-100.

Schaub, A. 2012. Aachen in römischer Zeit aus archäologischer Sicht – Versuch einer Neubewertung. In: Raban von Haehling/Andreas Schaub (Hrsg.), *Römisches Aachen. Archäologisch-historische Aspekte zu Aachen und der Euregio*: 131-205. Regensburg.

Schaub, A. 2008. Gedanken zur Siedlungskontinuität in Aachen zwischen römischer und karolingischer Zeit. *Bonner Jahrbuch* 208 (2008) 161-172.

is to be seen as an independent public latrine, whose location near the thermal complex 'Bücheltherme' was probably just occasioned by the large drains present there. It seems likely that both latrines of the 'Bücheltherme' also were accessible to the general public outside of the thermal complex. This is indicated by their respective peripheral location in the thermal complex. The possible latrine 3 was located directly to the east of the great central square (forum?) of *Aquae Granni* and must have been easily reached from there.

The accumulation of latrines in the area of the 'Bücheltherme' is without doubt a result of the state of research and does not reflect the real distribution of latrines through *Aquae Granni*. There has to have been at least one latrine in the 'Münsterthermen' thermal complex as well. At the moment we also know nothing about smaller, private installations. More will hopefully be known after the extensive documentation of the excavations of the last decades has been analysed.

(translated by Stefanie Hoss)

An outhouse in the garden? – Looking at a backyard in the *vicus* of Bonn

Jeanne-Nora Andrikopoulou, Manuel Fiedler and Constanze Höpken

Introduction

In 1989 – the year the city of Bonn celebrated its second millennium – the Rheinische Amt für Bodendenkmalpflege (Office for Archaeological Monument Conservation in the Rhineland) carried out an excavation lasting several months on the site of the (then) future museum 'Haus der Geschichte der Bundesrepublik Deutschland' in Bonn (Andrikopoulou-Strack 1990: 78–79; Andrikopoulou-Strack 1996: 421–468). The area researched lay in what was then the government district, and what in Roman times had been the *vicus* of Bonn. This had developed around the mid-1st century AD c. 2 km south of the Roman legionary fortress and its *canabae* (Figure 1). The settlement's main artery in Roman times, which ran almost identical to the modern Adenauerallee in a NS-direction, separated the *vicus* into two parts. According to earlier excavations, the settlement took up an area of around 60 ha, stretching to a length of c. 1,5 km. In the 6000 m² building pit to the west of the Adenauerallee, the rear parts of several Roman strip houses were excavated (Figure 1, excavation plan). As the modern street lies above the front end of the Roman houses, these could not be excavated. The houses had been erected in framework, were up to 40 m long and had been preserved to floor level at best. Some of the houses had had *hypocaust* heating and wall paintings.

Our knowledge of the structure of Roman *vici* in Germany has vastly improved during the last decades, mainly due to many large-scaled excavation. During these excavations, the attention was also directed at the back yards of the strip houses. Research like the ones carried out in *Lopodunum*/Ladenburg, *Arae Flaviae*/Rottweil or Heidenheim have given us broad insight into the use and structure of these areas behind the houses. They consistently were not built up, except for some smaller buildings, cellars and sheds. Instead, close rows of rectangular or round pits were discovered. In the *vici* of southern Germany, these sometimes were interpreted as cisterns, but first and foremost as latrines (Sommer 1992: 301).

The free spaces behind the houses seem to have been used as gardens and as parts of workshops of craftsmen in Bonn as well. In addition to several wells, many pits of different depths were found, which in part had been lined with wood. After their abandonment, they had been filled with rubbish.

Description of the latrine

On the plot of a strip-house that had been badly preserved due to modern building, a brick construction was discovered during the creation of the *planum*. The brick construction was located c. 12,5 m from the assumed rear wall of the house (Figure 1). As visible in the plan, the construction is made up of two almost square chambers (Figure 2). The section shows that the building was set into the natural ground, and that the brick-rubble walls reach different depths (Figure 3). A wall placed almost in the middle separated the building into a chamber and a deep shaft. The preserved overall length varies between 2,25 – 2,30 m, the preserved width between 1,05 m and 1,35 m. The western shaft was slightly larger than the eastern chamber. It enclosed an interior room of 1,10 x 0,90 m, the chamber on the other hand enclosed an area of only 0,75 – 0,80 x 0,85 – 0,95 m.

The walls – made from brick rubble in careful layers – vary in width: in the east, the width is between 0,10 and 0,15 m, in the west between 0,25 and 0,30 m. The levels also are different: In the east, the lower level was c. 0,15 m under the level of the planum and thus shallower than in the western half, where the long sides were 0,55 m deep, while the short sides were even 0,65 m deep on the western border of the context. The eastern chamber was preserved 0,64 m deep and filled with brown clay, in which darker inclusions, broken pottery and small pieces of stone and broken bricks lay. The western shaft on the other hand had been preserved to a depth of 1,60 m. It was filled with dark brown clay with green inclusions of sediment and small stones, pottery and an iron fitting, the whole filling being quite loose and penetrated by a lot of roots. In the lower half, bigger brick fragments were frequent. Like many other pits and wells in the excavation area of the 'Haus der Geschichte', the filling included the bones of several dogs.

The construction can be described in detail with the help of the section: In a first step, a pit with vertical walls and a flat bottom was dug into the natural ground (Figures 2 and 3). The bottom of the western half was about 0,16 m lower than the eastern side. From this

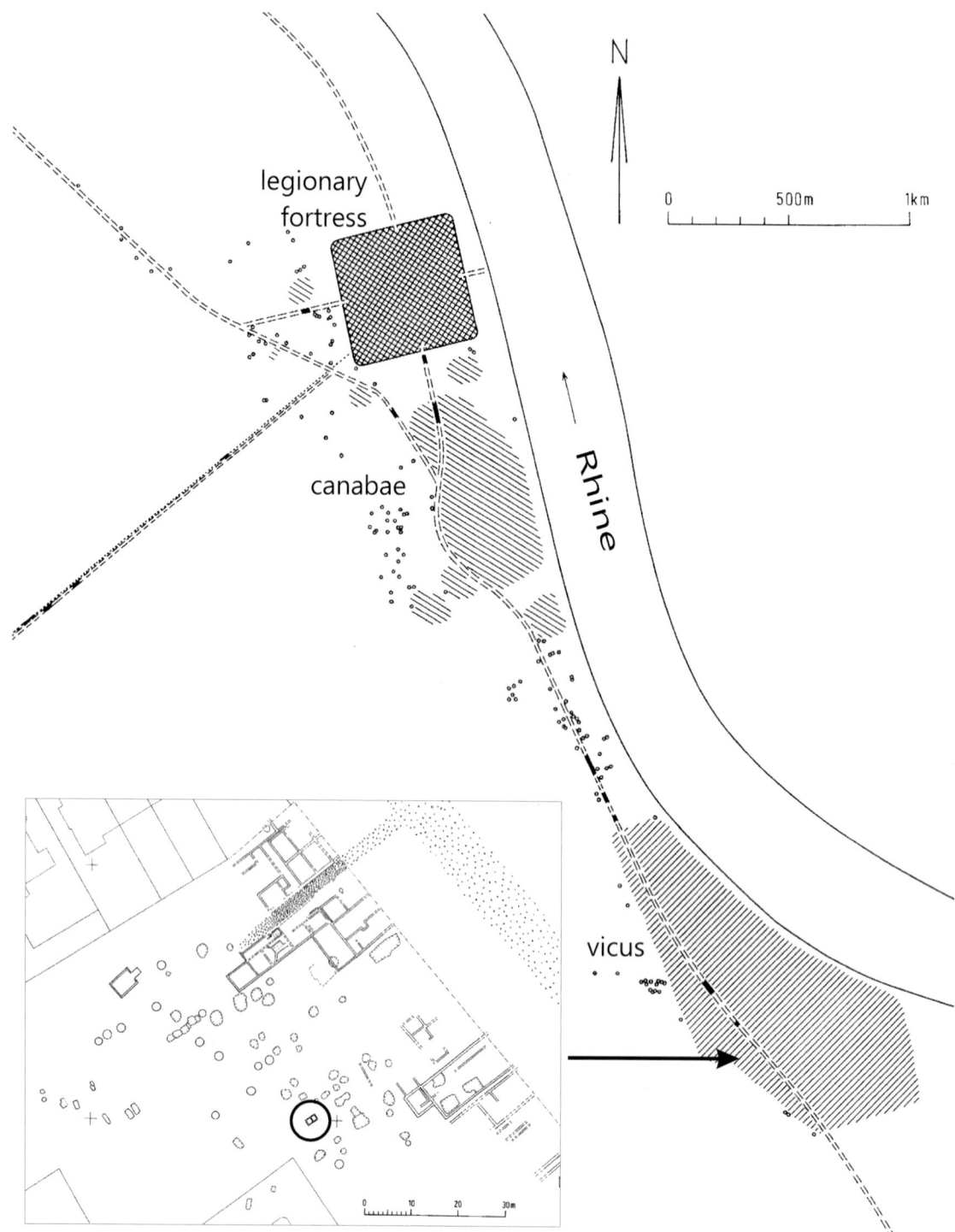

Figure 1. Location of the excavation area in the *vicus* of Bonn and location of the latrine in the excavation area (LVR-State Service for Archaeological Heritage in the Rhineland).

level, the western shaft was dug out a further c. 0,70 m, most likely to place an empty barrel in it. In the corners, the remains of decomposed wooden posts were preserved. The empty space between the shaft wall and the barrel or other wooden lining was filled with ground. The wall of the entire upper part of the pit had been dug out to different depths to introduce the brick-rubble foundations. Subsequently, a wall of large brick rubble was erected almost in the middle of the large pit, separating it into two chambers. In a last step, the eastern side of the building pit was filled up again.

This installation has no direct parallel. The shallow depth of the barrel, which did not reach the ground water table, excludes an interpretation as a well. Square or rectangular shafts lined with stone or brick and situated behind houses have been interpreted as possible cesspits in the *vicus* of Heidenheim (Rabold

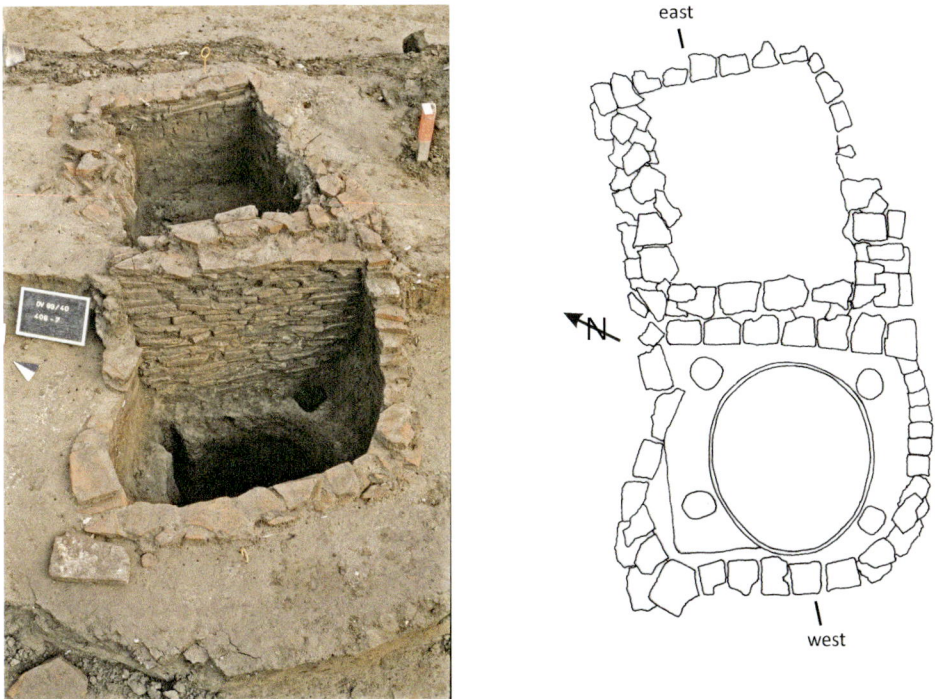

Figure 2. Plan of installation looking east (M 1:50, LVR-State Service for Archaeological Heritage in the Rhineland)

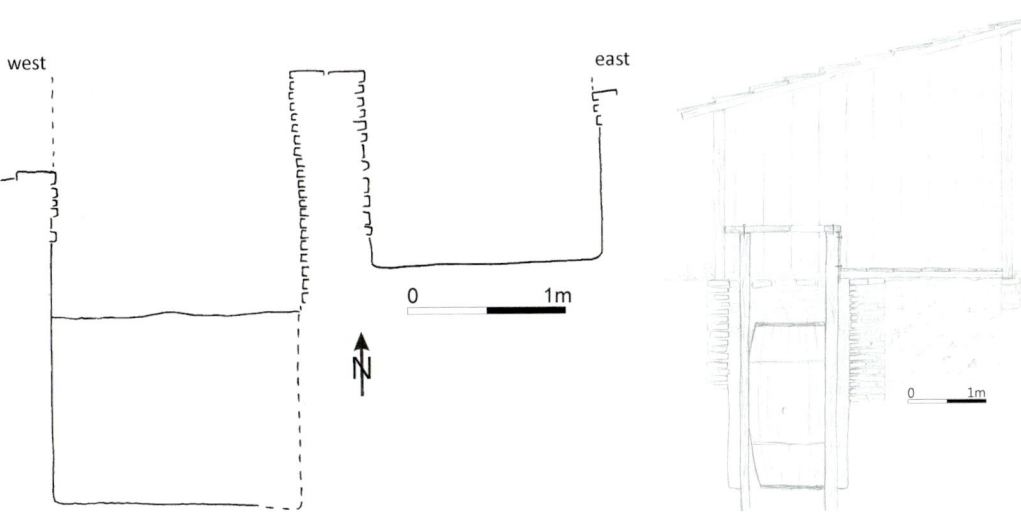

Figure 3. Northern section, M 1:50 (LVR-State Service for Archaeological Heritage in the Rhineland) and reconstruction of the installation M 1:100 (C. Höpken)

1993: 144–148, fig. 102). The square plan and the brick lining of the upper part of the installation in Bonn corresponds with those finds, but the latrine shafts of Heidenheim do not seem to have had a wooden lining or a barrel set into the deeper part of the shaft. However, a wooden lining with corner posts and a wooden floor has been found in a collapsed square pit with an original depth of 1,5 m in the *vicus* of Ladenburg (Kaiser and Sommer 1994: 77 Abb. 54–55).

The search for an appropriate parallel leads us to contemporary development aid for in the Third World, as the manner of construction of latrines without a water supply or drain has not changed in modern times. In developing countries, so-called cesspit-latrines are still being built, whose fundamental principles are similar to the installation in the *vicus* of Bonn. According to a relevant construction manual, the floor of a latrine should be higher than that of the surroundings. The superstructure is placed above a pit, which may be lined in the upper part and enforced in the lower part – in Bonn a barrel was used. A distance of at least 0,50 m should be kept between the floor and an assumed maximum filling level. This would have been the case in the Bonn latrine, if the barrel had originally reached up to the height of the brick lining, but also if it was shorter.

Comparable to the modern toilets, a wooden superstructure standing on the brick foundations can be reconstructed. The size of this shed presumably was the same as that of the brick foundation, ca. 2,20 m long and 1,30 m wide. The cabin opened towards the house; it was entered from the east. The western half of the cabin was occupied by the toilet itself; the eastern part was presumably equipped with a wooden floor. The necessary construction of a holed seat rested on the long posts, which had been placed in the shaft between the pit walls and the wooden barrel. The diameter of the barrel of ca. 0,8 m shows that this must have been an installation for the use of one person at a time.

If this interpretation of the building is correct, we are looking at an outhouse from the Roman period, corresponding in form to Bouet's 'fosse simple cuvelée' (see Bouet 2009: 26–33).

In difference to the inhabitants of the houses in Italian cities like Pompeii or Herculaneum, who had interior toilets, this outhouse in the garden vividly illustrates how the inhabitants of a strip house in the *vicus* of Bonn solved a prosaic daily problem.

The finds from the filling

The finds retrieved in four separate entities form a confirmation of the assumed construction sequence of the latrine. They had been extracted from the top layer, and from the eastern and the western shaft and finally from the foundations.

In the top layer of the filling, eight fairly complete vessels had been found, which provide a date for the last filling and the abandonment of the latrine (Figure 4,1–9). Apart from a loom weight and one coarse bowl, the pottery is all fine ware: Three slipped vessels and four Terra Sigillata vessels, with three of the latter being bowls of the types Drag. 37, Drag. 44 and Drag. 45 (Figure 4,1-3). The mortarium type Drag. 45 with a pouring spout in form of a lion's head emerged during the late 2nd century AD. The rim is slightly undercut, conforming to the development of the early 3rd century (Pferdehirt 1976: 58–62; Pirling and Siepen 2006: 67–68). The bowl Drag. 44 emerged during the mid-2nd century and was used until the end of the 3rd century – with a noticeably more perpendicular wall (Pferdehirt 1976: 54–55; Pirling and Siepen 2006: 65–66). The bowl with relief decoration Drag. 37 carries between the punches a stamp of Comitialis (Hartley *et al.* 2008: 95, stamp 5a tab., also see Ludowici 1905, pl. 256: Comitialis c). This company had workshops in Haute Yutz (F), Westerndorf, Trier and Rheinzabern (all D) between 160 and 240 AD (Hartley *et al.* 2008: 95–102). The relief ornaments are composed of the following stamps: warrior with lance M181a, warrior M212, running dog T139, pointed leaf without midrib P38, double leaf P127a, ovolo E23 (Ricken and Fischer 1963: 80. 93. 151, 192. 210. 301). The stamp and the punches belong to Comitialis' style IV – consequently this is a product of the workshop at Rheinzabern, produced after 180 AD (Hartley *et al.* 2008: 100, 102; Mees 2002: 335 (Comitialis IV), 336 (compare Comitialis V). The fourth Terra Sigillata vessel is a small dish type Drag. 40 (Figure 4,4), another type of the second half of the 2nd and the 3rd century AD (Pirling and Siepen 2006: 53).

The slipped beaker (Figure 4,5) probably is a product of a pottery in the Argonne region, as indicated by the orange fabric in combination with the matte black slip (for possible production centres see Vilvorder 1999). The type is common from the mid-3rd century onwards (Pirling and Siepen 2006: 82. 106). The bowls (Figure 4,6–7) are Rheinish products of the late 2nd and early to mid-3rd century (Höpken 2005: 88; Pirling and Siepen 2006: 115).

The coarse bowl with thickened rim (Figure 4,8) has been used for a long period of time from the 1st to the 4th century AD. The remains of soot on both the inside and the outside of the bowl are particularly interesting. Experiments have shown that these soot patterns appear when a vessel filled with a liquid is put next to the fire to heat the contents (Höpken 2011). Because of the clear demarcation upwards in this pot, which emerges when liquids are taken out and the pot is exposed to the fire again, the level of the contents can be deduced as well.

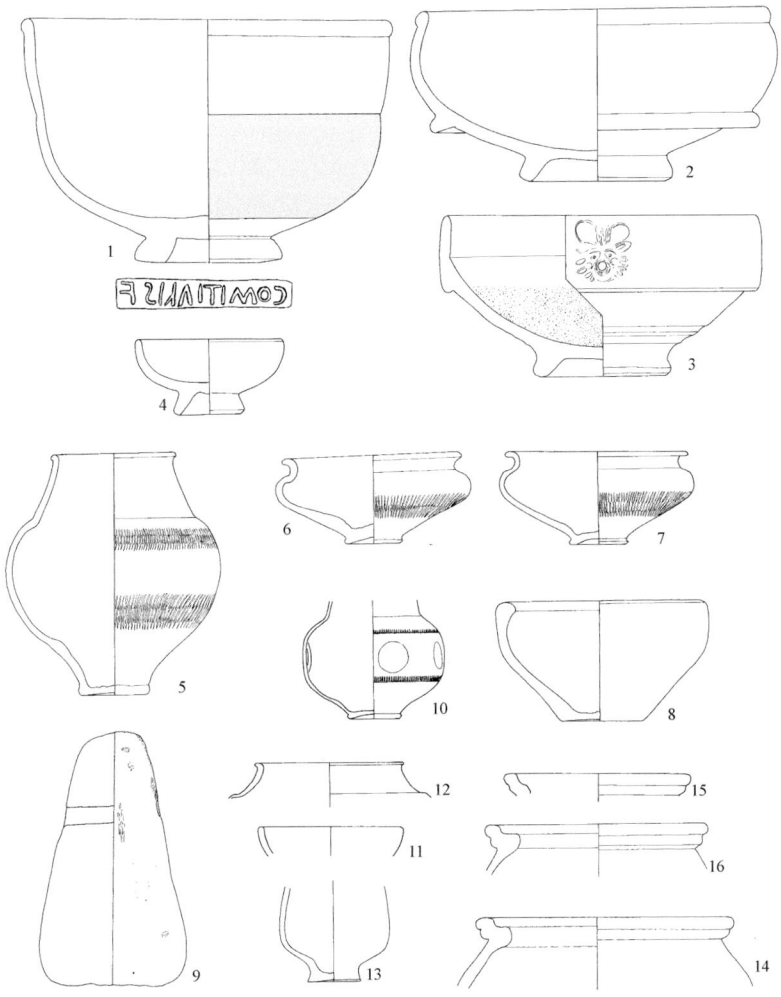

Figure 4. Pottery from the top filling (nr. 1-9) and from the western shaft (Nr. 10-16), M 1:3, stamp M 2:3 (Drawings C. Höpken).

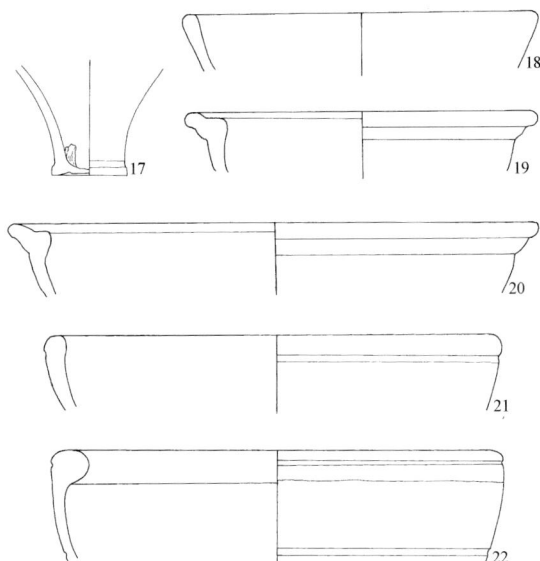

Figure 6. Pottery from the eastern chamber (Nr. 17-22, M 1:3, drawings C. Höpken).

The loom weight has a rectangular section and is coarsely made (Figure 4,9). A fragment of a second weight with a round section was also found. They can be included into the large group of loom weights found over the whole excavation area.

The upper parts of the filling also contained fragmented pottery. Parts of Terra Sigillata-dishes type Drag. 18/31 and 32, cups Drag. 33 und 40, and a mortarium type Drag. 43 were discovered, as well as fragments of numerous slipped vessels. Among them were several dishes, a face pot, several indented beakers and at least one exemplar each of the beaker forms Niederbieber 30 and 32. In addition to that, sherds of plain jugs were found – remarkable is a base fragment with a hole cut post cocturam and of unknown purpose. The coarse ware is represented by pots type Niederbieber 89, bowls with thickened rim and lid and mortaria. This broad spectrum fits in with the period around the mid-3rd century already indicated by the complete vessels.

Only a few pieces of coarse ware pots and red-slipped vessels – most of them locally or regionally produced – were found in the western shaft (Figure 4,10–16). They often have a mineral crust, which is the result of the conditions in the surrounding sediment. Not of Rhenish origin are a beaker from a workshop in Trier (Figure 4,10), which can be dated to the early 3rd century AD and a Terra Sigillata cup type Drag. 40 (Figure 4,11, Pirling/Siepen 2006: 79–80). Among the other finds, the fragments of coarse pots type Niederbieber 89 (Figure 4,14–16) dominate, a form which is well-represented in the pottery spectrum of the mid-2nd to 4th century AD (Höpken 2005: 127–128; Pirling and Siepen 2006: 225–226.). One of them has black incrustations from the last use. The plain ware is only represented with a few sherds. The fine ware comprises several types of beakers (Figure 4,12–13). The beakers Niederbieber 30 and Niederbieber 32 were in use from the second half/late 2nd century onwards (Höpken 2005: 79–80. 81–82; Pirling and Siepen 2006: 110–111. 113).

In the eastern shaft, fragments of mainly coarse ware were excavated, and some fragments of a plain jug (Figures 5,17–22, 6,23–31). Fine wares were barely represented, except for some body sherds of what seems to have been a beaker Niederbieber 32 and some body sherds of Terra Sigillata. The small diameter of the base and the shape of the plain jug (Figure 5,17) indicate a date no earlier than the late 2nd century AD (Höpken 2005: 108.). Among the coarse ware, fragments of pots type Niederbieber 89 were again found (Figure 6,27–31). In addition to that, a lid-seated bowl (Figure 5,19, Höpken 2005: 122) and bowls with (slightly) thickened rims (Figure 5,21–22), lids and a pan (Figure 5,18, 6,23, 6,26). The whole ensemble is to be dated in the 2nd century AD or later.

In the clay and brick foundations, only two plain and three coarse ware body fragments of Rhenish origin were found, which may be generally dated into the 2nd century AD.

The erection of the latrine can thus be dated into the 2nd century AD. The pottery finds from the eastern chamber seem to be slightly younger than those from the western shaft and the top layer and confirm the construction in the late 2nd century. The latrine was abandoned during the mid-3rd century, as confirmed by the finds from the top layer.

The choice and preservation of the pottery in the top layer are remarkable, as almost exclusively complete fine ware bowls and beakers were found. Generally, complete vessels are only buried in unusual circumstances, such as in layers of destruction or abandonment or as a positive selection in the context of grave goods or depositions. The fact that the vessels are mainly fine wares implies that this was a conscious choice. It cannot be excluded that they represent a ritual of closure on the abandonment of the installation, in which some vessels – probably with an organic content – were deposited. This is supported by another complex of vessels of very similar composition, found in a pit not far to the SW, which is thought to have been a latrine.

(translated by Stefanie Hoss)

Bibliography

Andrikopoulou-Strack, J.-N. 1990. *Der vicus von Bonn – ein römisches Dorf unter dem Regierungsviertel*, (Archaeologie im Rheinland 1989): 78–79. Köln.

Andrikopoulou-Strack, J.-N. 1996. Der römische vicus von Bonn. *Bonner Jahrbücher* 196: 421–468.

Hartley, B. R., Dickinson, B. M., Dannell, G. B., Fulford, M. G., Mees, A. W., Tyers, P. A. and Wilkinson R. H. 2008. *Names on Terra Sigillata. An Index of Makers' Stamps and Signatures on Gallo-Roman Terra Sigillata (Samian Ware)* (Institute of Classical Studies 102, Supplement 102-03). London.

Höpken, C. 2005. *Die römische Keramikproduktion in Köln.* (Kölner Forschungen 8). Mainz.

Höpken, C. 2011. Gebrauchsspuren an römischer Keramik. Beispiele aus dem Südvicus von Straubing. *Jahresberichte des Historischen Vereins für Straubing und Umgebung* 113: 41–70.

Jansen, G. C. M. 2000. Studying Roman Hygiene: the battle between the 'optimists' and the 'pessimists'. *Cura Aquarum in Sicilia. Proceedings of the tenth international Congress on the History of Water Management and Hydraulic Engineering in the Mediterranean Region. Syracuse, May 16 – 22, 1998.* (BABesch Supplement 6): 275–279.

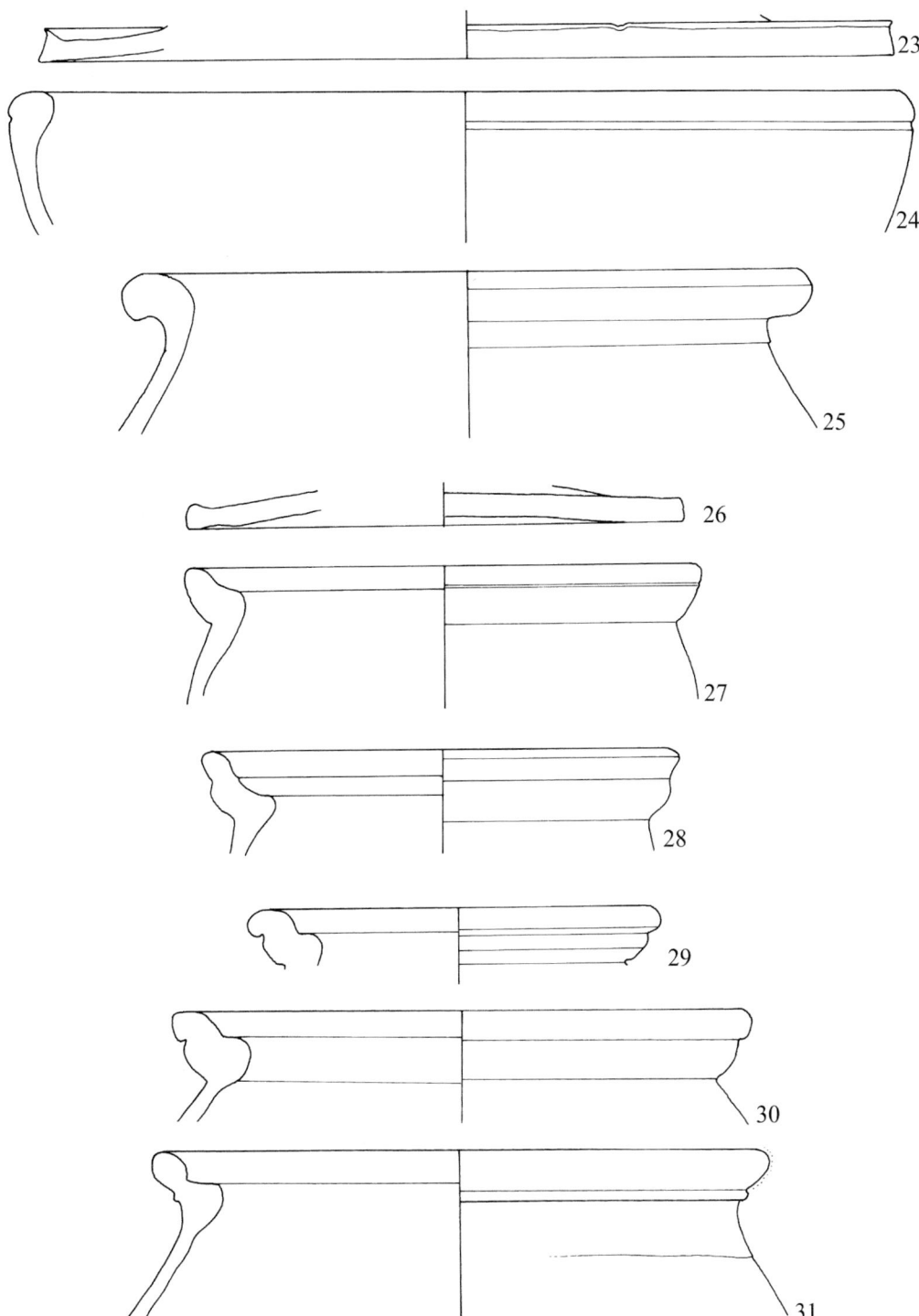

Figure 6. More pottery from the eastern chamber
(Nr. 23–31, M 1:3 drawings C. Höpken)

Kaiser, H. and Sommer, C. S. 1994. *Lopodunum I. Die römischen Befunde der Ausgrabungen an der Kellerei in Ladenburg 1981 - 1985 und 1990* (Forschungen und Berichte zur Vor- u. Frühgeschichte in Baden-Württemberg 50). Stuttgart.

Ludowici, W. 1905. *Stempel-Bilder römischer Töpfer: aus meinen Ausgrabungen in Rheinzabern, nebst dem II. Teil der Stempel-Namen 1901-1905*. Jockgrim.

Mees, A. W. 2002. *Organisationsformen römischer Töpfer-Manufakturen am Beispiel von Arezzo und Rheinzabern unter Berücksichtigung von Papyri, Inschriften und Rechtsquellen* (Monographien Römisch-Germanisches Zentralmuseum 52). Bonn.

Neudecker, R. 1994. *Die Pracht der Latrine. Zum Wandel öffentlicher Bedürfnisanstalten in der kaiserzeitlichen Stadt*. (Studien zur antiken Stadt 1. Bayerische

Akademie der Wissenschaften, Kommission zur Erfortschung des antiken Städtewesens). München.

Pferdehirt, B. 1976. *Die Keramik des Kastells Holzhausen.* (Limesforschungen 16). Berlin.

Pirling, R. and Siepen, M. 2006. *Die Funde aus den römischen Gräbern von Krefeld-Gellep* (Germanische Denkmäler der Völkerwanderungszeit B, Die fränkischen Altertümer des Rheinlandes 20). Stuttgart.

Rabold, B. 1993. *Einem römischen Handwerkerviertel auf der Spur. Ausgrabungen in der Heidenheimer Ploucquetstraße* (Archaeologische Ausgrabungen Baden-Württemberg 1992). Stuttgart.

Ricken, H. and Fischer, Ch. 1963. *Die Bilderschüsseln der römischen Töpfer von Rheinzabern.* Bonn.

Sommer, C. S. 1992. *Municipium Arae Flaviae – Militärisches und ziviles Zentrum im rechtsheinischen Obergermanien. Das römische Rottweil im Licht neuer Ausgrabungen.* (Berichte der Römisch-Germanischen Kommission 73). Mainz.

Vilvorder, F. 1999. Les productions de céramiques engobées et métallescentes dans l'Est de la France, la Rhénanie et la rive droite du Rhin. In R. Brulet, R. P. Symonds and F. Vilvorder, *Céramiques engobées et métallescentesgallo-romaines. Actes du colloque organisé à Louvain-la-Neuve le 18 mars 1995*: 69–122. Oxford.

A bath with public toilets in the *vicus* of Bonn

Gary White

During the excavation of the Roman *vicus* of Bonn (see also contribution Andrikopoulou-Strack *et al.* in this volume), a small Roman bath building was discovered. It was located in the eastern part of the Bonn *vicus* on the lower western terrace overlooking the Rhine. A road cutting passed by the baths on its way from the riverbank up to the *vicus*-plateau above. It formed a morphological dividing line between the steep impact slope of the meander to the north and the flat alluvial deposits to the south.

Both phases of the baths were comprised of a main building containing the bathing rooms, around which various subsidiary extensions had been added (Figure 1). The longitudinal axis of the central structure was SW-NE. In its layout, it conformed to a design in which the bathing rooms were arranged in a single row, i.e. the main building was heated at only one end with the bathing rooms being placed one behind the other according to the respective decrease in room temperature required.

The cold room (*frigidarium*) stood within the NE end of the central structure. This room was unheated and was also significantly shorter as either *tepidarium* or *caldarium*. Additionally, the room had been divided below the floor level into two unequal parts along the width. Whilst the NW section served as a square walking

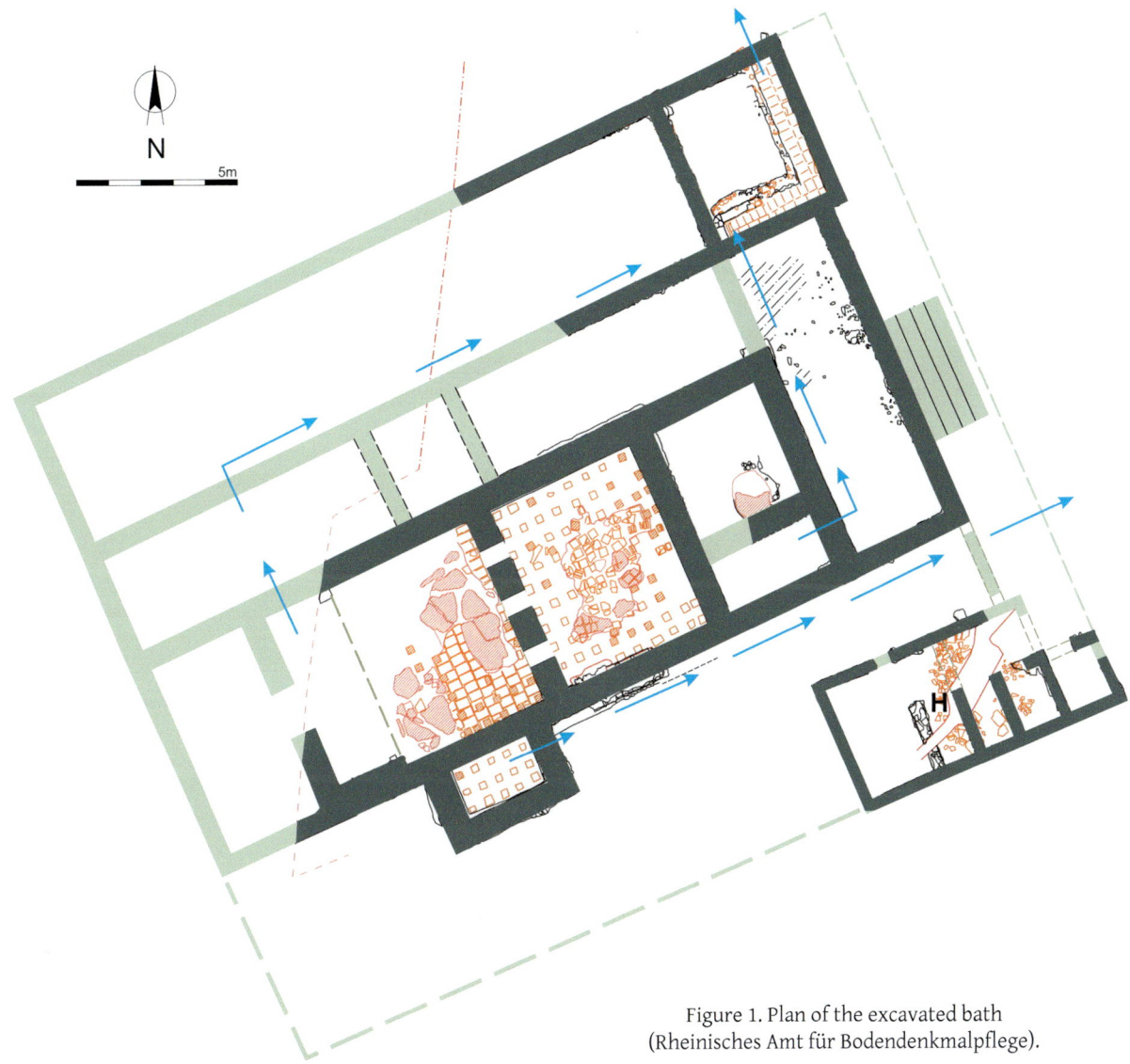

Figure 1. Plan of the excavated bath (Rheinisches Amt für Bodendenkmalpflege).

area, the smaller section to the SE of the dividing wall, being double in length than in width, had originally contained the cold water bath (*piscina*). Because of the divergence of 90° in the orientation of the cold room to the longitudinal axis of the structure, it can be assumed that one entered the bathing tract through the NW door of this room.

In the functional middle of the bathing tract was the *tepidarium*. It had a *hypocaust* under the entire floor. The floor, which had once been supported by tile pillars, lay caved-in at the centre of the room. The positions of most of the pillars were recognizable as faint, light-coloured impressions on the soot-stained surfaces of the floor and walls. Three openings in the dividing wall to the *caldarium* allowed the continuation of the soot-stained concrete floor of the *hypocaust*. These channels were the point at which the hot air currents entered the area under the warm room. As the functional centre in the main building, the *tepidarium* probably communicated with the other two bathing rooms as well as having had an exit to the entrance hall to the NW.

The last bathing room in the SW of the main building was the *caldarium*. The empty part of the room had the same dimensions as the *tepidarium*. The warm water basin (*alveus*) was placed in the extension to the SW along the whole width of the room.

About 3 m to the NW of the main building was a foundation running parallel to the outer wall. It was narrower as the walls of the bathing rooms and its foundations did not reach down as far, which indicates that they had a lighter structure to carry. If the entrance to the room was opposite of the *piscina*, the position of the extension between the street and the main building would have been the ideal place for the location of the entrance hall and an *apodyterium* (changing room).

The absence of underground channels, which could have taken the wastewater directly from the two basins to a drain in the street or to the main drainage system, was unusual. Similar to other comparable buildings one would have expected at least an opening in one of the foundation walls of the *frigidarium*, where a constant flow of water left the cold water basin. There was, however, no indication of a cut through the stonework or of the remains of an underground channel. The connection must have been at a higher level than the archaeological layers preserved in front of the *frigidarium*.

This was also the case with the wastewater disposal from the *caldarium*.

An L-shaped stretch of a channel that lay somewhat deeper was found on the NE corner of the baths. At the end of the long arm of the canal, which led towards NW, a pit was found that slightly undercut the end of the drain. During the first building phase, it must have functioned as a sump – perhaps with a filter basket of organic material – at the junction between the channel and the drain in the street, as proven by the many small finds that had gathered here. The length of channel coming at right angles from SW was walled-up at the starting point. This termination of the channel was not a later amendment, but had been built at the same time as the sidewalls. Apparently, the deeper part of the channel had been planned from the outset to begin here. It was striking, that the outer face of the blocked end of the channel was in a line with the back wall of the *frigidarium*. In addition, the long front wall of the entrance tract ended at the termination of the channel. Thus the underground part of the channel seems to have begun at the corner of the original building. Perhaps the rainwater from the roof was collected at this corner and directed into the drainage canal.

The second phase of the baths required modifications in the area of the *caldarium* and *praefurnium* as well as the addition of large hall and public toilets in the northwest. A new entrance area with a flight of steps to the east bears witness to the re-orientation of the whole structure by 90 degrees in the direction of the Rhine. Finally a two-storey service building with a broad entrance to the courtyard was placed at the SE corner.

In the *caldarium*, the *alveus*, which was at the end of the room and directly heated from the *praefurnium*, was broken out and renewed at a higher level. At first, the floor of the *praefurnium* was raised by means of in-filling and then the firing canal under the bath was also brought up to the required level. The basin could be emptied using the former channel. Raising the level of the *praefurnium* floor as well as adding steps at the other end of the building, was very likely the result of terracing the slope and laying down new streets.

NE of the main building and entrance hall was a rectangular annex with the same internal width as the former entrance tract. While the annex used the already existing walls to the west, a new L-shaped stone foundation provided the outer walls to the east. The foundations were set against the drainage channel and the main building, showing very clear building seams.

To the northwest of the original bath building, a long, rectangular extension had been added. Here, either existing structures, such as the outer side wall of the former entrance tract and the drainage channel, were integrated, or new foundations were built out of re-used material from the main building. The extension had been divided by at least one wall.

If we assume that the southwest section of the annex was not divided up into smaller units, the room would

Figure 2. Toilet during the excavation (Rheinisches Amt für Bodendenkmalpflege).

have been big enough to allow an interpretation as a *basilica thermarum*. Such hall-like amenities served as meeting points where physical exercises, amusements and sociable pastimes took place.

At the north-eastern end of the building extension there was a small room. The drainage channel, which during the first phase was outside of the building, was now completely integrated within the annex. The eastern walls used the outer sidewall of the channel as a foundation so that the channel ran along two walls inside the room (Figure 2). The earlier filtration pit at the end of the canal was given up and the water was directed further away. The new room can with confidence be interpreted as a toilet. The individual seating facilities would have been above the channel along the two adjacent sides of the room.

The location at the junction of two important streets indicates that the toilet was intended as a public amenity for the use of the pedestrians and people about their business on the road. It would have been possible to enter the room from the side road north of the building as well as from the *basilica thermarum* to the west. Such toilets were nearly always to be found built onto the large urban bathhouses because only they offered a reliable supply of water for flushing the channel.

The provision of public toilets for the Roman *vicus* in Bonn aptly fits a period of expansion and remodelling in the development of the town and illustrates the upgrading in the urbanisation and the self-confidence of its citizens.

The dating of the second building phase of the baths can only be roughly ascertained. An analysis of the coins found in a levelled layer of burnt material under the service building resulted in an earliest possible date for the redevelopment in the first half of the second century. How close this *terminus post quem* lies to the actual building period is not known. According to the finds within a wooden structure – which ceased to exist when the baths were remodelled – it would seem that the second half of the second century is the most likely date of renovation for the erection of the public toilets.

The Roman public toilet of Rottenburg am Neckar

Stefanie Hoss

Introduction

This article is meant as a short introduction to the Roman public toilet of Rottenburg am Neckar. Unfortunately it cannot be more, as the whole complex – excavated more than twenty-five years ago by now – is still virtually unpublished. At present, the only publications available are a guidebook on Roman Rottenburg and a preliminary report in the publication of the department of archaeology of the state of Baden-Württemberg plus a catalogue entry in Bouet's book on public toilets (Bouet 2009; Heiligmann 1992; Reim 1987).

Fortunately, the impression given by the publications can be augmented by visiting the toilet, which has been preserved on site in the building that was the cause for the excavation– a multi-storied parking garage. In addition to exhibiting the Roman toilet in *situ*, the ground floor also functions as the Roman museum of *Sumelocenna*.

The poor state of publication is especially frustrating considering that the Roman public toilet of Rottenburg is exceptional for three reasons: it is the only Roman public toilet on the right side of the Rhine (the so-called *agri decumates*), it is one of the rare cases where a Roman public toilet is not connected to a bathhouse and it is one of the largest and most luxurious installations of its kind in Germany.

Rottenburg – Sumelocenna

Today's Rottenburg is a small town about 50 km southwest of Stuttgart and a mere 12 km from Tübingen (Figure 1). It's forerunner, Roman Rottenburg was called *Sumelocenna* and was founded around 85-90 AD, probably in connection with a military fort (Heiligmann 1992: 23). Several inscriptions indicate, that by the time of Trajan – at the latest – *Sumelocenna* and the surrounding region were turned into a *saltus*, an imperial estate administered by an equestrian *procurator*, whose official residence was *Sumelocenna* itself. The reason for this is not known, but *Sumelocenna* is the only imperial domain on the right side of the Rhine (Heiligmann 1992: 34).

Sumelocenna lay at the crossing of several important Roman roads, the most important being the long-distance east-western trade route connecting eastern Bavaria with Switzerland (Figure 1, Heiligmann 1992: 24-27). The inscriptions discovered in the town are evidence for its regional importance as an administrative and political centre of the middle Neckar region. Under Marcus Aurelius, *Sumelocenna* seems to have become the central place of the *civitas Sumelocennensis*, which was bordered in the north by the *civitas Aureila G[...]* (probably modern Bad Canstatt), in the east by the provincial border with Raetia, in the south by the *Municipium Arae Flaviae*, modern Rottweil, and in the west by the *civitas Aureila Aquensis*, modern Baden-Baden (Heiligmann 1992: 34-36). It seems likely that this 'upgrading' in rank was the cause for a building boom in *Sumelocenna*, with many new buildings – some of them, like the public toilet, quite elaborate – being erected in the latter part of the 2nd century AD.

The Roman buildings of *Sumelocenna* excavated to this date include a temple precinct with two temples, three bathhouses, the public toilet and several larger houses, parts of the city walls and some pottery kilns outside the city walls (Gairhos 2008; Heiligmann 1992: 24-27).

Like many other places on the right side of the Rhine, Roman Rottenburg was probably destroyed in 259-260 during the incursions of the Alemanni. After this, the *agri decumates* region to the right side of the Rhine was given up by the Roman Empire. *Sumelocenna* seems to have been abandoned by is inhabitants and left to decay (Heiligmann 1992: 36).

Excavation area Sprollstrasse

During the years 1986-1991, a fairly large part of ancient *Sumelocenna* (2000 m^2) was excavated along the Sprollstrasse to enable the building of the new Hotel Martinshof and a multi-storied parking garage (Heiligmann 1992: 38). The area excavated lies in what today is the centre of Rottenburg, near the bishop's palace.

In this excavation area, a main street 4,5 m wide and running southwest to northeast was discovered, with a narrow side street running at right angles to it (Figure 2, the following after Heiligmann 1992: 39, 67-71). The *insulae* on both sides of this side street were only partly excavated but showed evidence for subdivisions into individual plots of c. 11 m wide and c. 40 m long. To the north of the side street, two of the strip houses typical for *vici* were partly excavated on plot I and plot II. The houses had been built from stone and covered about 250 m^2. Because of the bad preservation here, the inner division in rooms is difficult to reconstruct, but

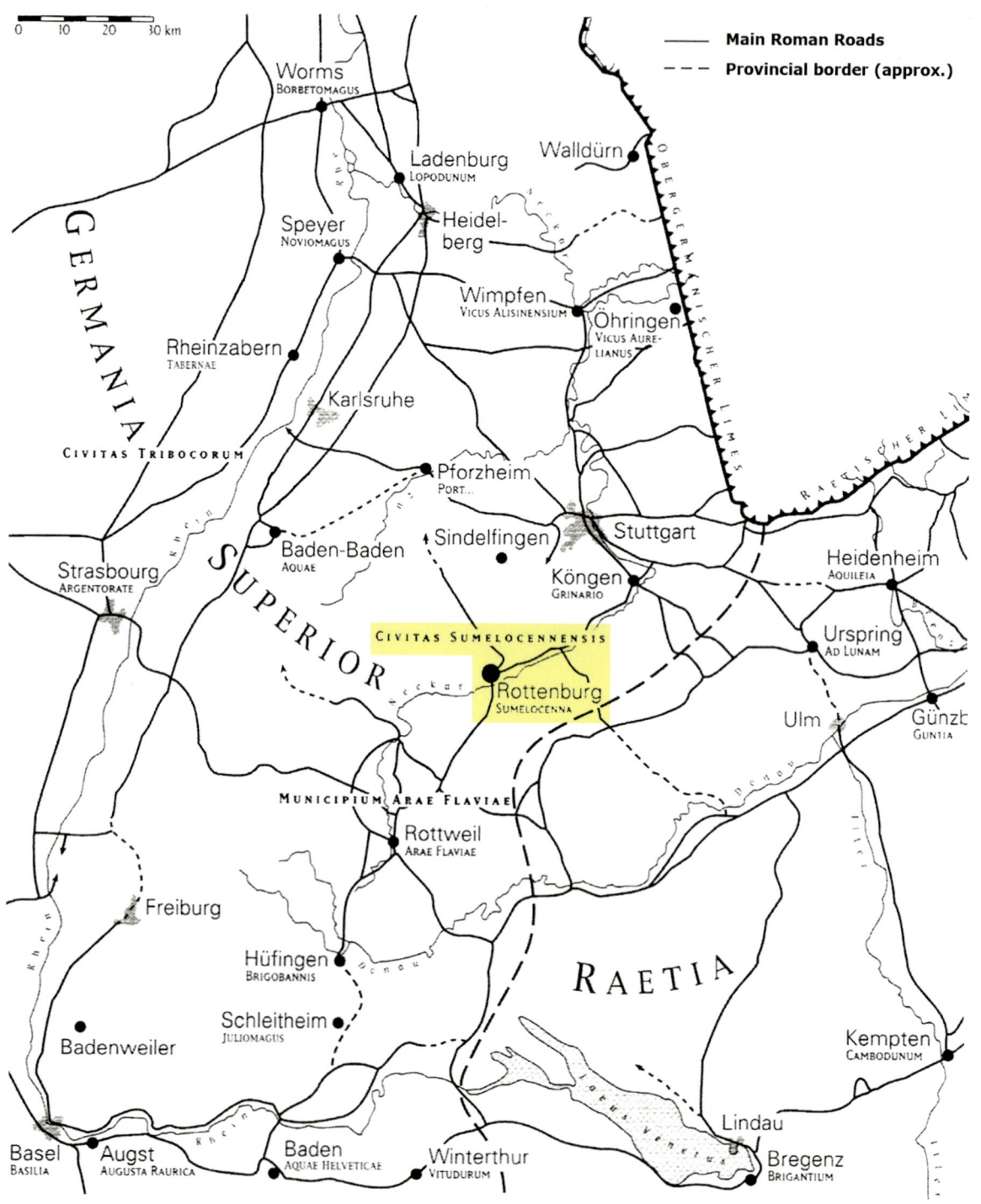

Figure 1. Map of the important roads in Southern Germany in the Roman period (after Heiligmann 1992).

both had a room with a hypocaust of around 13 m² and cellars. While the cellar of one house was situated in the front part of the house, the other house had two cellars in the rear part of the house, an uncommon placing. The side street was bordered in the north by the house in plot II and in the south by the public toilet. Street and toilet together took up the space of plot III.

To the south of the toilet lay plot IV, with a four-roomed house of 140 m². The two front rooms were fronted by a *porticus* and accessible from the street. One of the rooms had a channel *hyopcaust* and the other a hearth. The house had a cellar with an airshaft in its eastern wall in the backyard. Its floor was of stamped mud and still had indentations of amphorae in it. The foundations of the building and the fact that it sits about 1 m higher than its neighbours clearly show that it had been build later and filled an empty space between the public toilet and the large house with a *peristyle* to the south. The house was most likely used as a shop or workshop, perhaps a *taberna*.

The large house occupying plot V had several phases and was only partly excavated. It was most likely planned symmetrically around a *peristlye* with a cross-shaped pool and wall paintings. This would imply that it also took up the neighbouring plot, which was not excavated. It would then measure 50 x 24 m or 1200 m² and be very large indeed. At least two rooms had *hypocaust* heating, one of them also having an apse. The evidence points towards this house as having been quite prestigious.

Public toilet

The public toilet was a long and narrow building sitting on top of the main sewer of the city, which was 80 cm wide and lay 1 m underneath the public toilet. This sewer first ran under the main street, but curved in a wide bend to run under the centre of the public toilet (Figures 2, 3, the following after Heiligmann 1992: 53-56). The public toilet itself was 5,30 m wide and excavated to a length of 32 m without reaching the rear wall of the building in the east.

The entrance from the street lies on the western side of the public toilet and is 1,60 m wide. The angled staircase creates a small vestibule, after which 11 well-worn steps of local sandstone lead about 1 m downwards into the public toilet proper (Figure 4). This feature – an angled entrance situation – can be found in many public toilets as it prevents passers-by on the street to look into the toilet (personal comm. G. Jansen, 16 May 2011).

The room was divided in the longitudinal axis by at least seven sandstone columns with a height of 3 m. They had originally been plastered white as proven by some remains. The columns stood 4 m apart and slightly to one side of the central axis, so that they were placed above the southern sewer wall rather than directly above the sewer itself. As the room was very narrow, the columns are not structurally necessary. They must have been placed to monumentalize the building.

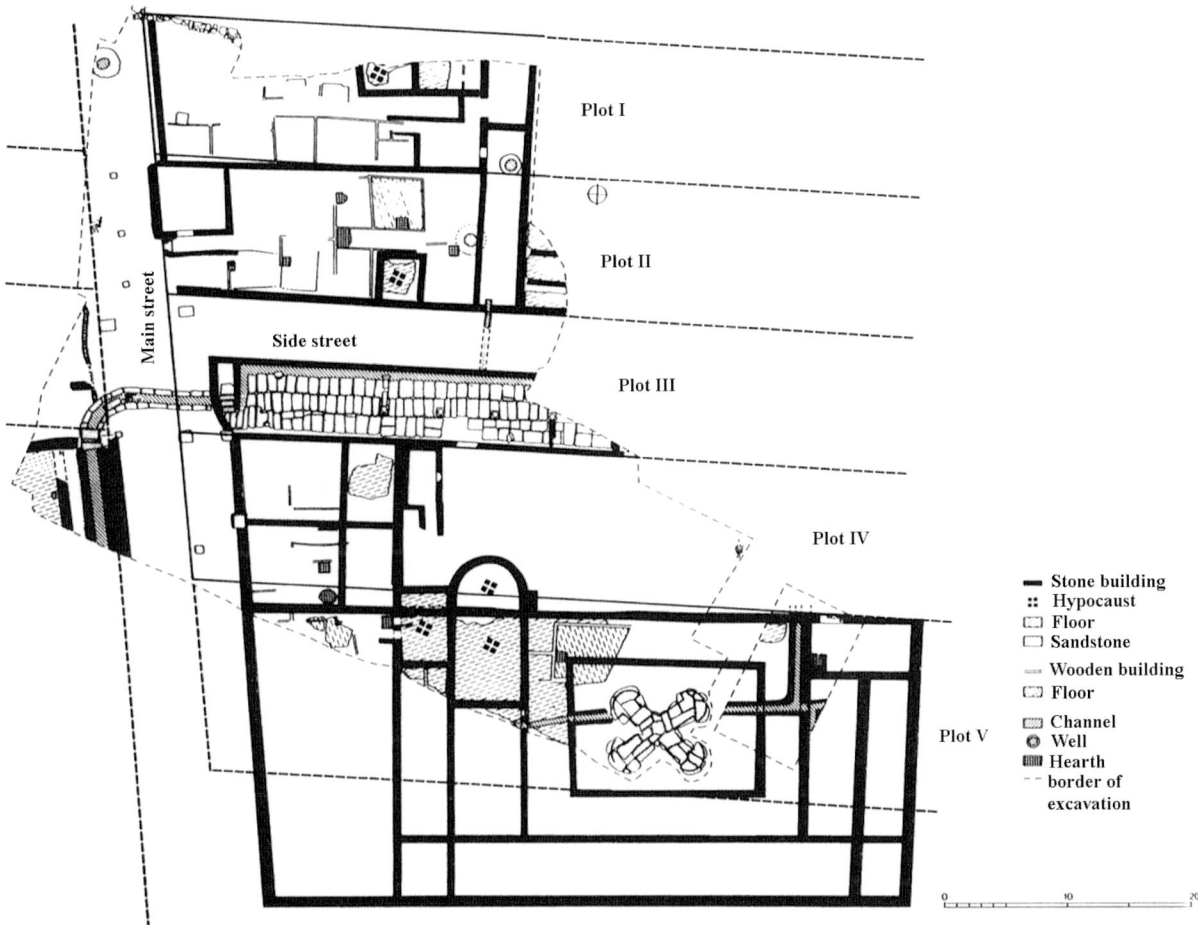

Figure 2. Plan of the excavations (after Heiligmann 1992).

Figure 3. Photo of the toilet after excavation, looking east (after Heiligmann 1992).

Figure 4. Photo of the toilet during excavation, looking west at the entrance (after Heiligmann 1992).

On the northern wall of the room ran a channel 50 cm wide and 90 cm deep with a brick floor, filled with water from a branch of the main city sewer running down the middle of the room under the flagstones (Figure 5). Heiligmann speculates that some sort of sluice regulated the flow of the water from the main city sewer to the branch channel.

Above this channel, the wooden planks forming the toilet seats had originally lain. The dimensions of the seats must have been 60 cm in depth at a height of 40 cm. Another small and shallow channel ran in the flagstones in front of the seats (Figure 6).

The whole room was laid with flagstones of the local sandstone. As was noted with the stairs, this type of stone wears quite easily which can also be seen on several places before the latrine channel. The worn places in the flagstones are evenly spaced and attest that the holes in the seating plank above the channel must have been placed about 90 cm apart, making it possible for at least 35 persons to use the public toilet at the same time. Alain Bouet even reconstructs up to 54 seats, as he also includes the short western side and the likely eastern part of the public toilet where the southern wall is preserved much longer than the northern wall (Figure 7, Bouet 2009: 360-361).

Figure 5. Toilet channel (after Heiligmann 1992).

Figure 6. Photo of the toilet during excavation, looking west at the entrance (after Heiligmann 1992).

A shallow surface channel in the flagstones ran along the southern wall (the following after Heiligmann 1992: 56-57). It is interpreted as carrying fresh water to a basin found in the room. This basin was also made from the local sandstone. It was round and had a scooped-out bowl of 30 cm depth (Figure 8). As the basin has one straight outer side, it is likely that it stood against the southern wall of the public toilet, perhaps at a spot about 20 m from the entrance. It may have been used for hand washing. Washing basins are known from public toilets in Ampurias (E), Dougga (TN), Housesteads (GB), Milet (TR) and several places in France (Bouet 2009: fig. 112).

Two foundations running across the flagstones from one long wall towards the other were observed in the western part of the building. This could point to a second phase in which a part or parts of the toilet were partitioned off – perhaps into a male and a female section (Heiligmann 1992: 57-58).

Both the northern and southern walls of the public toilet were decorated with wall painting (the following after Heiligmann 1992: 58-59; Bouet 2009: 140). The remains of the northern wall have been reconstructed as showing a roughly plastered base in red, which lay behind the wooden seats. Directly above the seats, a more finely plastered white zone with a height of 60 cm was decorated with yellow and black garlands, bound with thin red string. Above this was a zone with a series of horizontal stripes of varying width in red, white, sand and brown (Figure 9).

Above that, a zone of at least 1,80 m height was painted with floral patterns in yellow, purple and green on a white background. The windows must have been placed in or above the latter zone. Contrary to the northern wall, no picture of the southern wall was published in the descriptions available, but it is temptingly mentioned that the Roman visitors had left many graffiti (Heiligmann 1992: 49). The interior arrangement and design of the public toilet is dated to the late 2nd century AD (Bouet 2009: 360-361; Heiligmann 1992: 55).

Bouet has calculated that the dimensions of the Rottenburg public toilet are equal to that of the public toilet on the Southern Market of Milet and the Great Forica at the Largo Argentina at Rome. While the former also has a basin, the latter is also divided with columns along the longitudinal axis (Bouet 2009: fig. 30).

Contrary to Mediterranean public toilets of a similar size, the Rottenburg public toilet seems not to have been decorated with marble, mosaics and statues, as far as we know. These decorations are unusual in public toilets to the north of the Alps. In part, this may have been occasioned by practical reasons: Marble is not only quite expensive, but also prone to shatter when water and frost come together. Thus it surely is not an ideal material in unheated rooms where water is sloshed about even during frost in winter. The same can be said of mosaics, which welt and break in wet and freezing conditions. Stone also is an unusual material for toilet seats north of the Alps. In addition to the

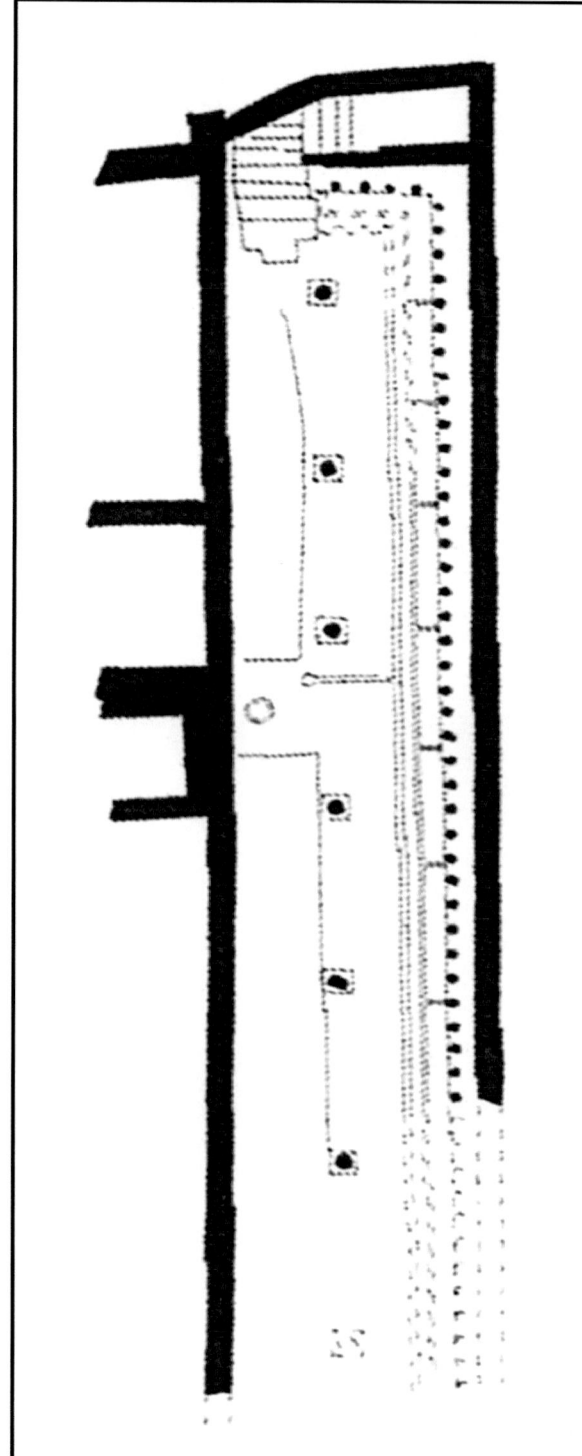

Figure 7. Drawn reconstruction plan of the toilet (after Bouet 2009)

Figure 8. Photo of the washing basin in situ (after Heiligmann 1992).

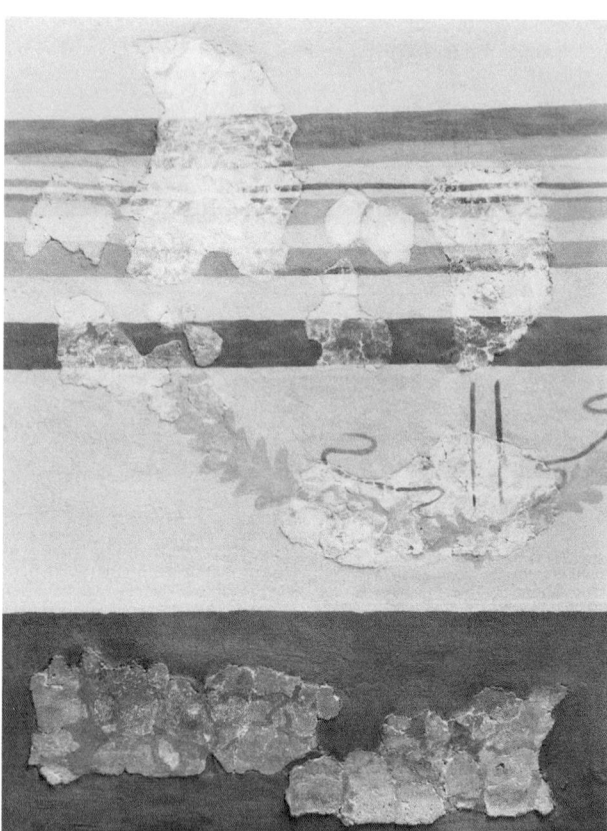

Figure 9. Photo of part of the reconstructed wall painting of the northern wall (after Heiligmann 1992).

aforementioned combination of frost and water, stone is always quite cold to the touch, and even more so on a cold day. Wood was probably the preferred material for toilet seats for this reason. This use of wood for toilet seats is not confined to the provinces to the north of the Alps though, it was quite common in Italy as well (personal comm. G. Jansen, 16 May 2011).

Although the decoration of the Rottenburg public toilet with columns, wall paintings and a stone basin might not be luxurious by Mediterranean standards, by the standards of a town in the Germanic provinces it was quite grandly fitted out. The dimensions of the toilet however are very impressive even when compared to Mediterranean examples, as the comparisons with two of the largest toilets of the Roman Empire, the public toilet on the Southern Market of Milet and the Great Forica at the Largo Argentina at Rome demonstrate. It must have been a very expensive building, proudly demonstrating the status of *Sumelocenna* as a wealthy town of importance.

Bibliography

Bouet, A., 2009. *Les latrines dans les provinces gauloises, germaniques et alpines* (Supplement Gallia 59). Paris.

Gairhos, S. 2008. *Stadtmauer und Tempelbezirk von Sumelocenna. Die Ausgrabungen 1995 - 99 in Rottenburg am Neckar, Flur 'Am Burggraben'*. Stuttgart.

Heiligmann, K. 1992. *Sumelocenna - Römisches Stadtmuseum Rottenburg am Neckar* (Führer zu archäologischen Denkmälern in Baden-Württemberg 18). Stuttgart.

Reim, H. 1987. Neue Ausgrabungen im römischen Sumelocenna, Rottenburg a. N., Kreis Tübingen. *Archäologische Ausgrabungen in Baden-Württemberg* 1987: 128-133.

Latrines connected to bathhouses in Germania inferior – an overview

Michael Dodt

Introduction

When looking for latrines with a channel flushing system, it makes sense to look at buildings producing a lot of 'grey' wastewater, namely bath buildings. In his handbook-cum-catalogue, Alain Bouet lists three such installations connected to bathhouses for the province of Germania Inferior: firstly the latrines in the two public bathhouses in Colonia Ulpia Traiana/Xanten, the third belongs to a private bath in Köln-Braunsfeld (Bouet 2009: 372–387). The latrines of the fortress of Oberaden (Bouet 2009: 373–383) topographically do not belong to the province Germania inferior, but to the short-lived province Germania (which ceased to exist when the provinces Germania Inferior and Superior were created under Domitianus round 89 AD). The entry on the latrines at the public bathhouse in Xanten in Bouet's book can now be supplemented with a few technical details which were published only recently by Norbert Zieling (see Zieling in this volume, Zieling 2008: 381, fig. 240-241). While these are unquestionably prominent examples, it is possible to find more latrines

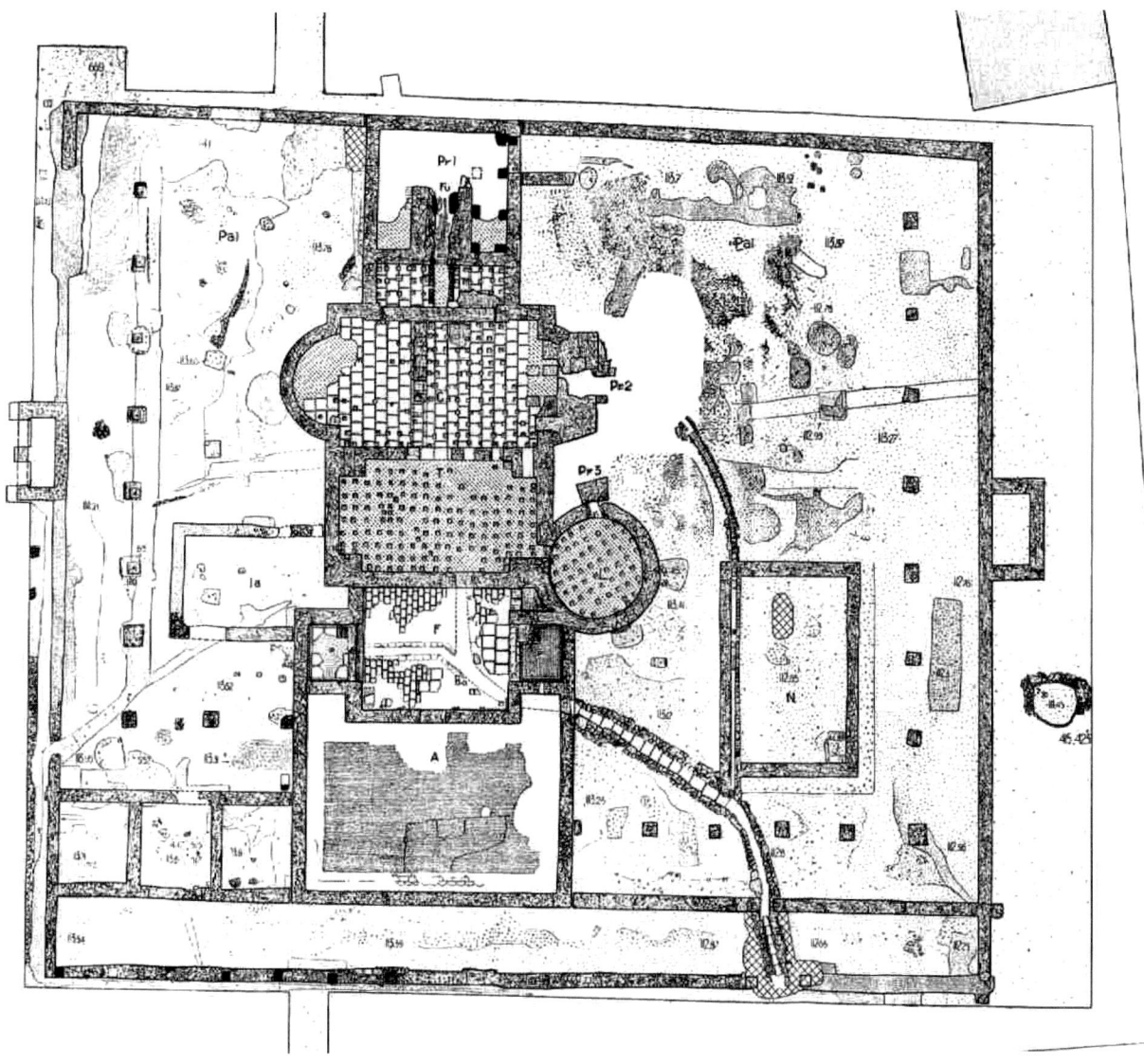

Figure 1. Thermae of Heerlen, excavation plan 1940 (after Van Giffen 1948, Taf. 3).

when looking more intensively at bath buildings in this province, just as Bouet has done for the Trier region using the article of Harald Koethe (1940) on villa baths in the Trier region. This contribution is an attempt to fill the gaps left by Bouet. While it developed out of the author's current research on baths and villa baths in Germania Inferior (Dodt 2003, Dodt 2006), it makes no claim to completeness. The research area Germania Inferior until now excluded baths on the territory of the Tungri, as it follows the older research on the provincial border between Germania Inferior and Gallia Belgica, in which the Tungrian territory is seen as being in Gallia Belgica.

Latrines at public bathhouses

1) *Thermae of Heerlen* (Provincie Limburg/NL) (Figure 1a/b)

The public bathhouse of Heerlen was build around the middle of the 1st century AD, excavated completely in 1940 and preserved for visitors in a large museum hall (Bechert 1982: 151-152; Bouet 2003: 679-681; Dodt 2003: 161-169, fig.. 85-87; Garbrecht/Manderscheid 1994, B, 183 -185 Nr. D 71-74; Heinz 1983: 80-82, fig.. 74. 79; Jamar 1981; Jamar 1988; Nielsen 1993, II, 21, Nr. C 154; Van Giffen/Glasbergen 1948).

The baths are described by Bouet as a building with a latrine under a building 'wrongly interpreted as

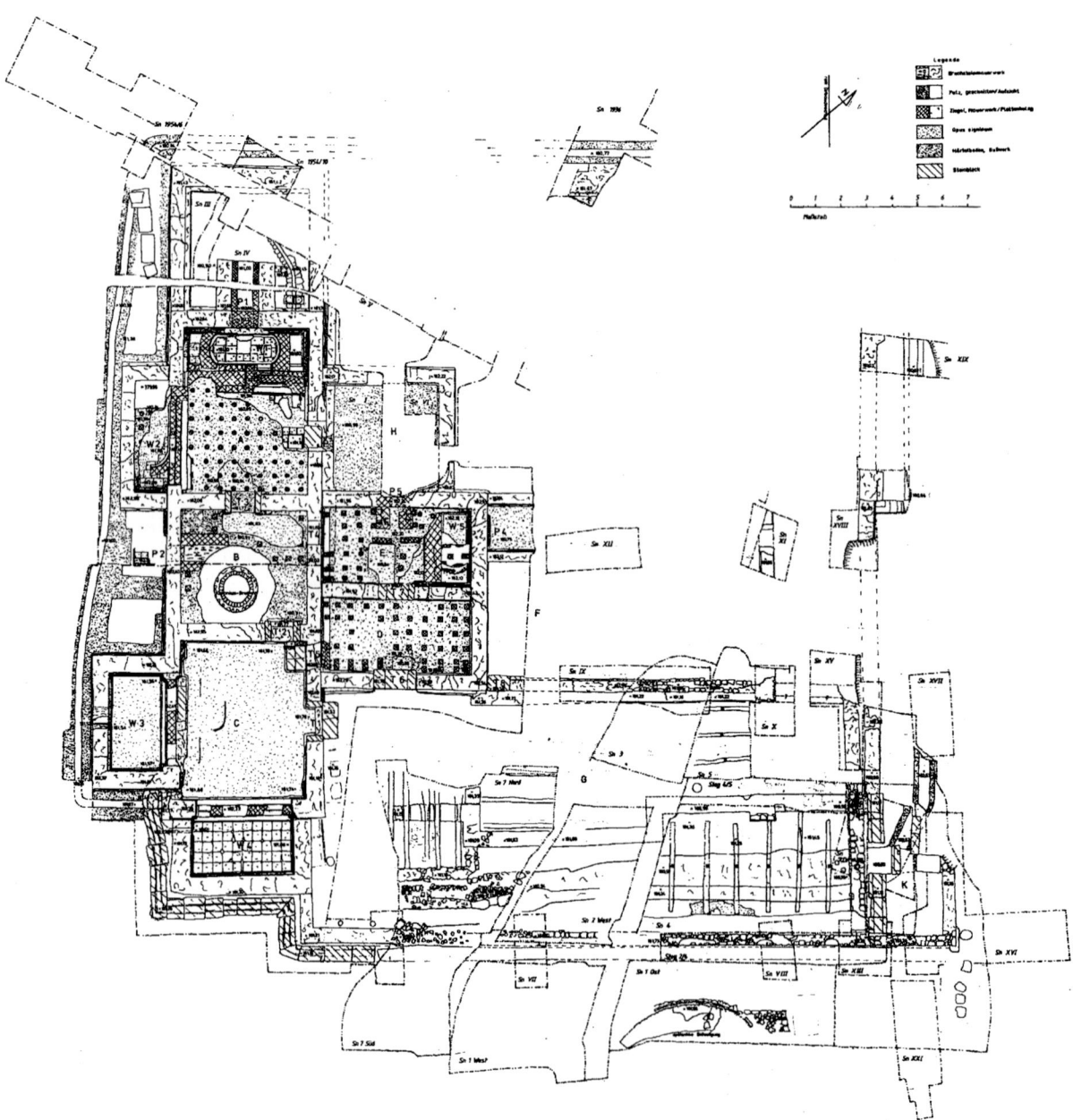

Figure 2a. thermae of Zülpich, excavation plan until 1979 (after Dodt 2003, Beilage).

a latrine' (Bouet 2009: 417-418. Nr. 245). In this, he refers to room Ia in the eastern *palaestra*, which had been interpreted as a latrine by the excavators (Van Giffen/Glasbergen 1948: 213). This room was built in the second building phase. An open drain running though the room and out of it has only been preserved as discolouration (The eastern *piscina* of the *frigidarium* is the only possible water source). Bouet points out that the room has none of the characteristics of a latrine, especially no channel along the walls.

A different room next to the entrance can be interpreted more clearly as a latrine because of its channel. This had already been built in the first building phase and received the wastewater from the eastern *piscina*, channelling it under the floor of the *frigidarium*, in which three more holes to the channel were later cut, to take up the spray water. West of the *frigidarium*, the channel received the wastewater of the western *piscina* and ran towards the corner of the *natatio* in the *palaestra*, where it met another channel from the *alvei*. All of the wastewater was led from there under the *porticus* of the *palaestra* and finally into the street-facing vestibule west of the entrance portal. Jamar assumes a separate room here, used as a latrine with the help of the channel, which empties into the Geelenbach creek 500 m west of the bathhouse (Jamar 1988: 30, 34).

The interpretation as a latrine arises from the course of the channel through the room. According to the documentation of the excavation and in the current state presented in the museum, the preservation of the context is very faint (Figure1). The channel also changes its direction after having broken through the rear walls of the porticoes at right angles. An interpretation of this room as latrine is thus very tentative.

2) *Thermae of Zülpich* (Nordrhein-Westfalen/D) (Figure 2a–c)

The small public bathhouse of Zülpich, whose main parts were integrated into a new museum in 2007, only has a single room with a channel running though it in the eastern corner of the bathhouse (Figure 2a). During all building phases, this channel received water from most pools. The only exception is pool (W5) of the later building phase III, which has an outlet to the *palaestra*, through which a channel ran in an eastwestern direction Contrary to the Heerlen bathhouse (no.1), the channel mainly runs on the outside of the bathhouse and finally through the aforementioned room in the eastern corner of the bathhouse. These are heavily destroyed, in contrast to the well-preserved main bathing rooms on the SW side of the bathhouse.

During the extensive rebuilding taking place during the approximately 200 years the bathhouse was in operation (mid-2nd to mid-4th century AD), the shape

Figure 2b. thermae of Zülpich, channels and walls at the east corner of the bathhouse area 1978 (after U. Heimberg/M. Gechter, Rheinisches Amt für Bodendenkmalpflege Bonn).

of the latrine was also altered (Dodt 2008: 107–109). One possible latrine room could only be attested for the last (third) building phase, dated to the end of the 3rd century AD (K, see Figure 2b). The room was more than 2 m wide, build from stone and had a channel running through it. It lay in the NE axis of the *basilica thermarum* and had a channel made from large sandstone ashlars running along the outer wall of the building. The channel started at the deep *frigidarium* pool W4 and ran along the outer wall of this room and the adjoining *basilica thermarum* G. The walls of the latrine have been completely removed in the NW, so that the length of the room is unknown. In the second building phase, a channel whose bottom and sides were made from mortar mixed with crushed bricks ran through the room (Figure 2c). This channel did not run along the NE wall of room G, but parallel to and about 2 m away from it, on the inside of the NE wall of the later room K. This is an indication that during the previous (second) building phase, a room must have occupied this space, which later vanished completely. It was connected to the eastern corner of the *porticus*, later replaced by the *basilica thermarum*. As the NE walls of the *porticus* and *basilica* did not lie on top of each other, the latrine of the second building phase can only be identified by this channel. The channel of the third building phase had been placed above the dismantled NE wall of the *porticus*. Unusually, the channel was not fed by the wastewater of the pools, but started at a rectangular basin on the

Figure 2c. thermae of Zülpich, channels and walls at the east corner of the bathhouse area (Photo Dodt 2002).

SE side of the *porticus*, next to pool W4, which has been interpreted as a cistern. In both the second and third building phases, the length of the latrine could not be determined, so the number of seats remains unknown for both building phases.

3) *Bath of the pottery district of Soller* (Nordrhein-Westfalen/D)
(Figure 3)

The pottery district of Soller (south of Düren) was discovered in 1932 during deforestation. An excavation took place in 1933 with enthusiastic private help under the direction of the local archaeologist/historic preservation officer of Düren, M. Bös. The results were published by Dorothea Haupt (1984: 391–470). Her fig. 4 (on page 399) shows the only plan, which was the basis for this analysis of the latrine. Even though the documentation of the excavation in the pottery district is preserved only in part, the interpretation of the latrine at the small bathhouse is clear. Although the bathhouse consists of a series of four rooms arranged in a row (a–d), they were not arranged in the conventional row type, as the *frigidarium* (b) and *caldarium* (c) lie in the middle (Haupt 1984: 398–400 (esp. 398, footnote 13), pl. 161,1). Together with the following room (a) to the SW, these rooms form a structure on a single axis, while room (d) on the NE was built slightly out of this axis and is wider than the rest. Just like *caldarium* (c), this room has a *hypocaust* and thus can be interpreted as a *tepidarium*, changing the usual order of rooms (Dodt 2003: 192). This is similar to the military bath at Donstetten (Heinz 1979: 43–45 pl. 2).

The floors of room (a) and frigidarium (b), were no longer preserved at the time of the excavation, but their walls were build in bonded masonry. A drain ran between these rooms, joining a channel a few meters to the SW. This channel drained the wastewater from the *piscina* of the *frigidarium* (Haupt 1984: 398-399, pl. 163 and 168,2). This channel is joined by another from the NE, which runs along the NW wall of the bathhouse at a distance of 1 m. Both the beginning and the end of this collecting channel have not been excavated. The probable endpoint of the channel, the Ellebach, runs in NE direction (Haupt 1984: 392–394, fig. 1–2).

The drain between rooms (a) and (b) is comparable to the latrine XVI of the *Villa rustica* of Lürken (our No. 6), so that room (a) can also be interpreted as latrine (Dodt 2003: 192). The water for the latrine may have come from the *alveus* of the *caldarium* (c), as this lies at the SE side of the caldarium – just like the start of the drain in latrine (a) – while there is no connection to the channel on the NW side.

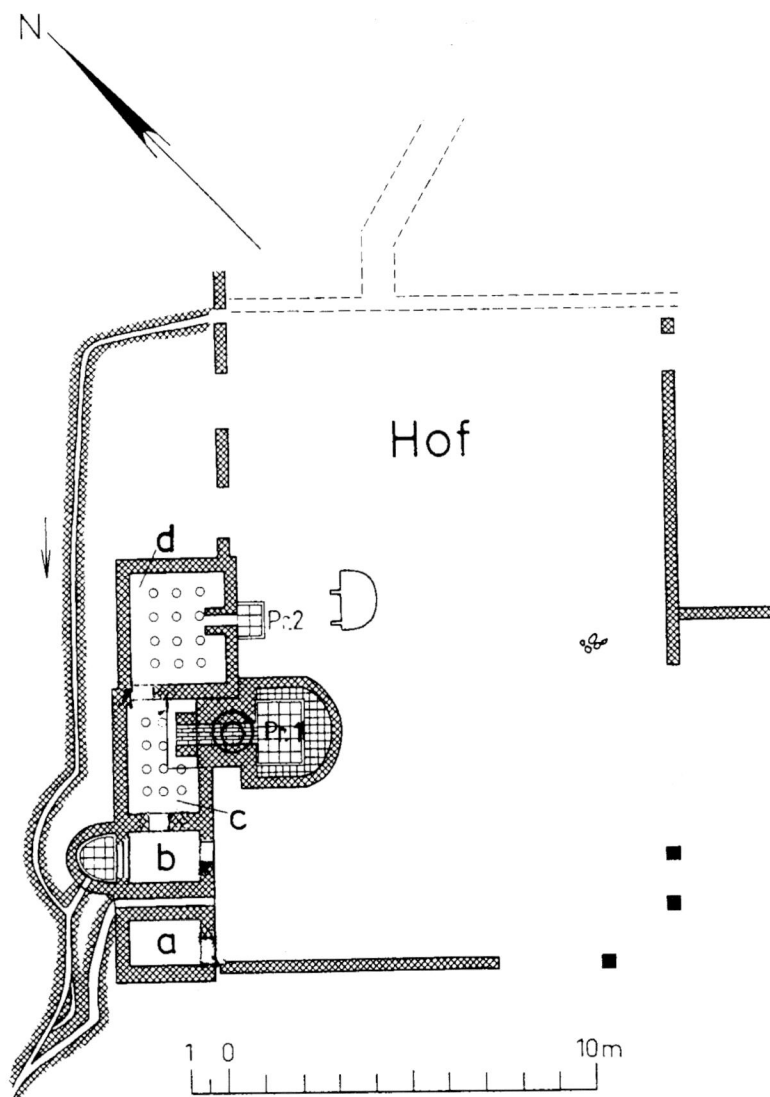

Figure 3. Bathhouse at the pottery district of Soller (after Haupt 1984: 399, Fig. 4).

4) Bücheltermen, Aachen (Nordrhein-Westfalen/D) (Figure 4a/b)

The so-called Bücheltermen are a Roman bathing establishment exploiting the hot, sulphurous Kaiserquelle spring (Cüppers 1982: 37–53. See also contribution Schaub in this volume). In the year 1958, during maintenance work at the sewage channels in front of the houses Büchel 32-36, just east of the Bücheltermen, a channel build from sandstone ashlars with a width of 0,60 m and a depth of 0,95 m was discovered (Cüppers 1982: 53–55; Hugot 1959: 276–280). The channel displayed formations of calcareous sinter, which make it likely that it drained the water from the 'swimming pool' (Hugot 1959: 377). The channel was connected to a long rectangular building (44) with a channel running along the inside walls. During excavation, only the narrow eastern and western sides and the NE and NW corners could be recorded. The walls between these points were projected to have run along straight lines. The entrance into the latrine may have been from this side, as it was in the latrine of the large public bathhouse at Xanten (Bouet 2009: 386, fig. 375 and contribution Zieling in this volume).

Latrine 44 had been added to the north end of the existing channel in a later phase, with older Roman buildings being torn down for it (Figure 4a, Cüppers 1982: 54-55; Hugot 1959: 379-380, fig. 24, pl. 52). Cüppers assumes an older latrine in the Roman walls M1 and M2 and the floor E1. Finds from the channel hint at an abandonment of the latrine in the 3rd century AD.

The building surrounded by walls M5, M7 and M8 had a full length of almost 19,50 m and a width of c. 5 m. Channel K1/K2 used for the transportation of the faeces had a width of 0,58 m – two Roman feet – between the walls M5 and M4, respectively M7 and M6. Its bottom lies at 158 m above NN, and thus 0,60 m above the level of the main channel, which would either indicate that

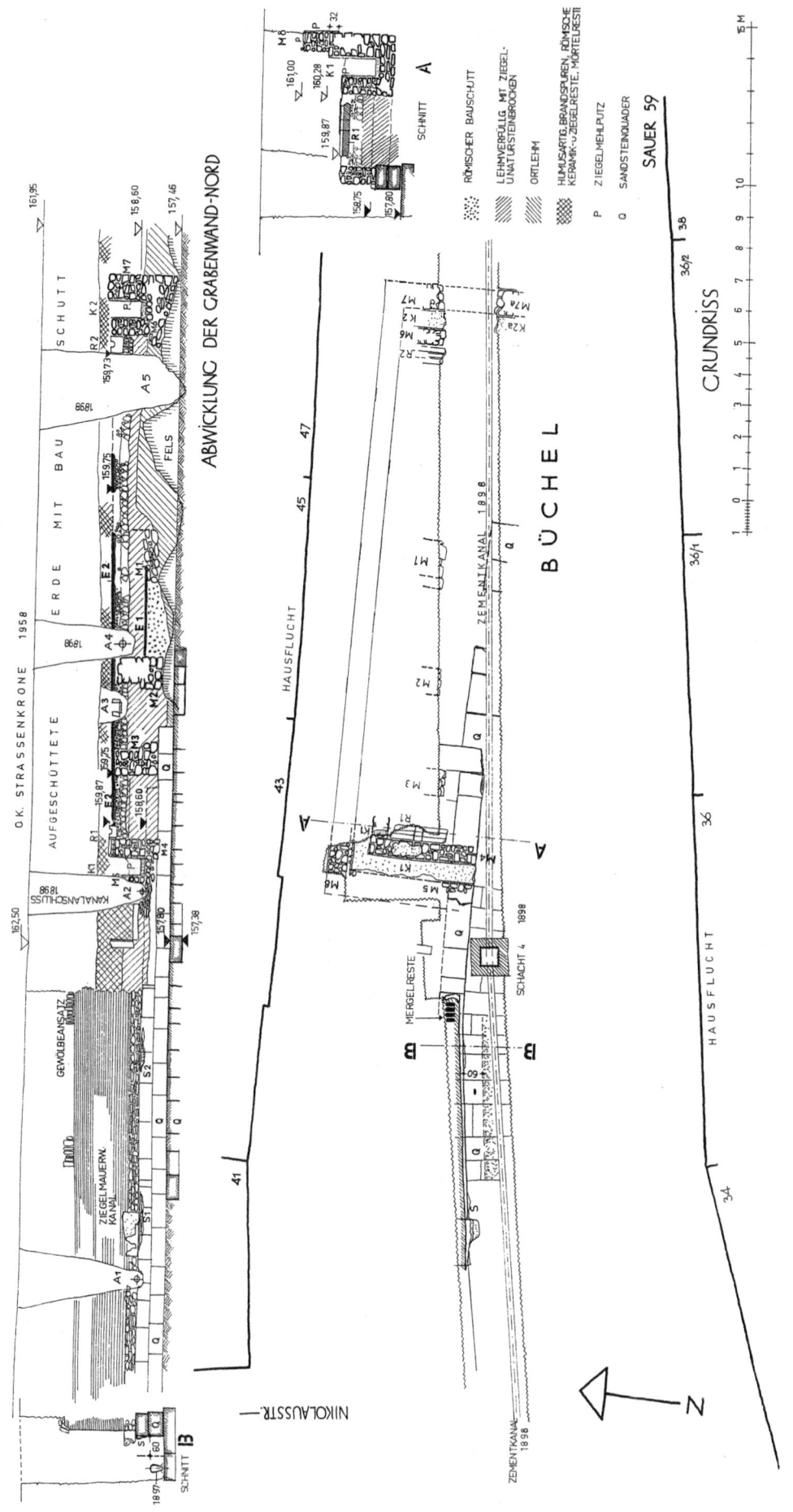

Figure 4a. Latrine at the Büchelthermen in Aachen, plan and section (after Hugot 1959, Taf. 52 and Fig. 24).

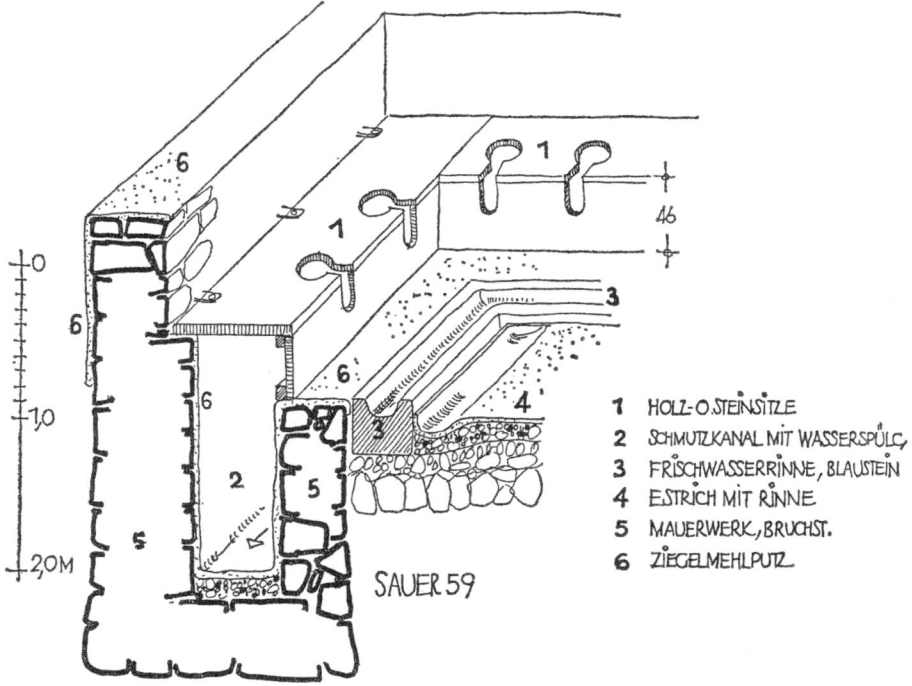

Figure 4b. Reconstruction latrine at the Bücheltermen in Aachen
(after Hugot 1959, Taf. 52 and Fig. 24).

channel K1/K2 could only be flushed during a high water levels in the main channel or – more likely – that flushing with wastewater was regulated with sluice gates as in the latrine of the public baths of Xanten (see contribution Zieling in this volume). A channel (R1/R2) made from Aachen bluestone (a blueish limestone found in Aachen) was running in front of walls M4 and M6. Its upper edge had the same level as the floor (Figure 4b), and it was probably used for the washing of hands. The floor (E2) was made with crushed brick and smoothed against the bluestone-channel, forming another little channel of 0,30 m width and 0,10 m depth, which may have functioned as a collector of wastewater during the cleaning of the floor. The walls are erected from *opus caementicium* and plastered with a 2 cm-thick plaster containing crushed bricks. A recess with a width of 0,12 m exists on the inner face of the outer walls M5 und M7, placed 0,41 m above the upper edge of the inner walls M4 and M6. This recess was used as a support for the wooden board forming the seat bench, while the perpendicular walls of the seat bench rested on the walls M4 und M6. If we assume the wooden board to have had a thickness of 4 cm, we can reconstruct a sitting height of 0,45 m (Hugot 1959: 379). The whole length of the seat benches on all three sides excavated is c. 22 m, which according to stone seat benches of similar lengths preserved in the Forum baths of Ostia or Vaison-la-Romain (Bouet 2009: 97, fig. 70) corresponds to around 40 seats. Another comparison is the reconstruction of the latrine of Lürken (our no. 6). If we also assume seats over the main channel, the number of seats for the whole latrine raises to 60, even when we factor in the door openings.

Latrines at private baths of villae rusticae

5) *Villa rustica of Ahrweiler* (Rheinland-Pfalz/D) (Figure 5a/b)

A less than well-preserved latrine with a water flushing system was found during the fairly recent excavation of the villa rustica of Ahrweiler-Silberberg (Dodt 2003: 214–220; Fehr 1993). The latrine was established during the second phase of the younger bath, which was erected round the end of the 1st century (no. 41; Figure 5a). The latrine lies in the higher part of the bathhouse, which – like the main building – was build on a slope. Consequently, the channel draining the *piscina* of *frigidarium* 31 and the *alveus* of *caldarium* 36, which both lay lower than the latrine, could not have flushed the latrine. The water must have come from the drainage channel running to the NE of the bathhouse, between it and the main building. This was probably built to ensure that the latter would not be undercut by water running off the slope. The entrance to latrine 41 is through a corridor (42) on the NE rear side of the bath, while the main bathing rooms have the classical orientation to the SW. The bath and the latrine had no direct connection, ensuring that the bather was not troubled by odours. The latrine itself is in the form of an L (Figure 5b), with a bottom of tiles and inner facings of the wall made from brick, while the outer facings of the walls were

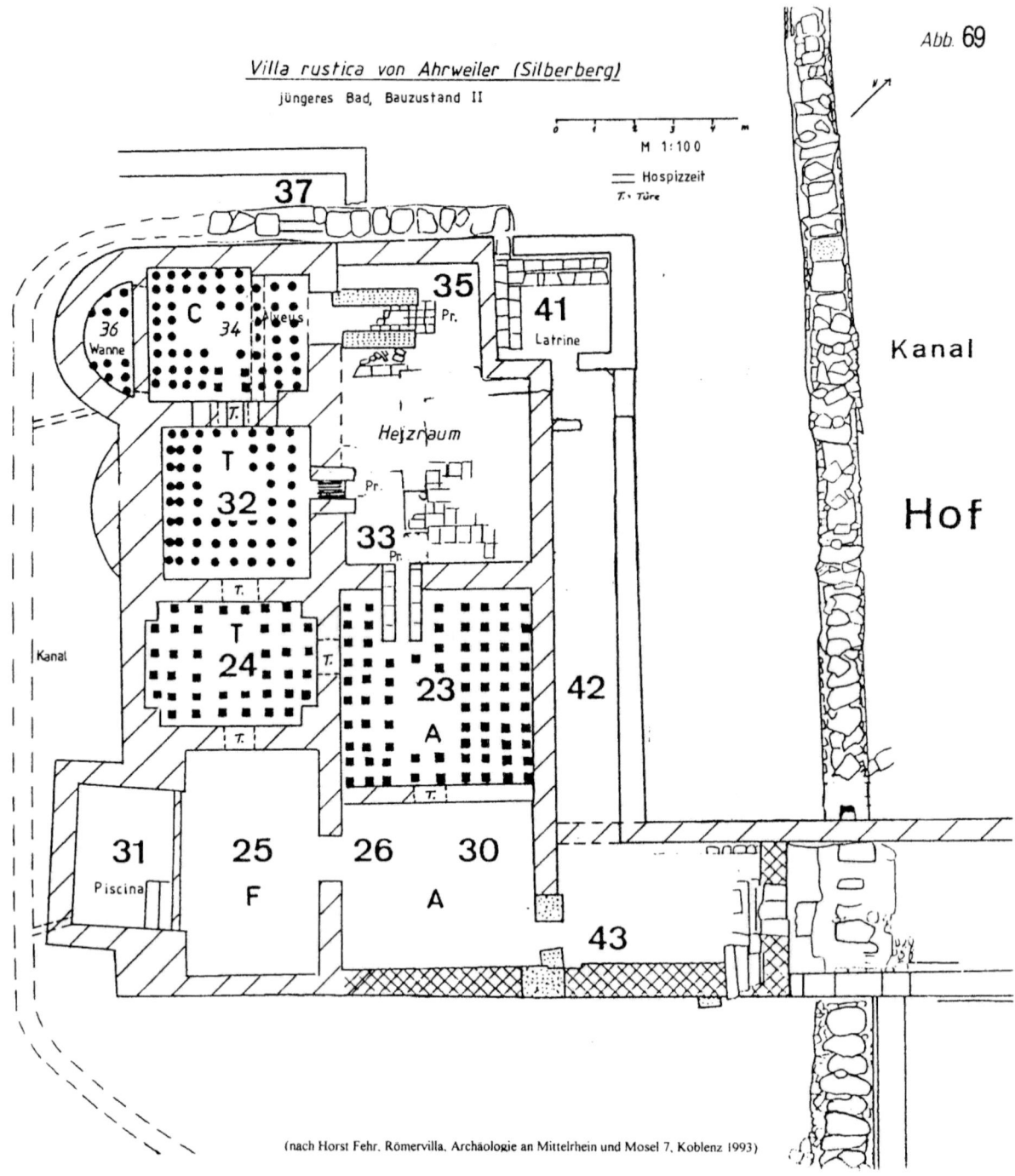

Figure 5a. Villa rustica of Ahrweiler, late bath at the SW side in the later building phase (Dodt after Fehr 1993, Plan 3).

made from greywacke and slate. On the western corner of room 41 and at an angle to the L-shaped channel lies the channel drain, which emptied into a rubble-build channel running on the NW side of the furnace room and the *caldarium*. It probably then ran around the *caldarium* to the southwest, to take up the wastewater from the *alveus* 36 and *piscina* 31 and drain them into the Ahr river, thus –unusually – draining the latrine first and then the pools. Unfortunately, this area could not be carefully documented during the 'strip, map and sample' rescue excavation and this conclusion is thus very tentative. As the latrine was drained first and the pools only later, the latrine could not have received the flushing water from them. It seems possible that both the latrine and the pools received their water through the drainage of surface water via a water-lifting well (I am indebted for this idea to D. Stender, Krefeld).

Figure 5b. Villa rustica of Ahrweiler, latrine at the northern corner of the bath, seen from the west (after Fehr 1993: 81 Fig. 46).

6) *Villa rustica* of *Eschweiler-Lürken* (Nordrhein-Westfalen/D)
(Figure 6a–c)

The extravagantly appointed bathhouse of the villa rustica von Lürken had two latrines (Figure 6a, Dodt 2003: 223–228; Kretzschmer 1981: 51–73; Piepers 1981: 34-35, pl. 20–26). The villa's bath lay at the rear part of the building and had been slightly altered during its use (Piepers 1981: 28, Figure 7, rooms IV, IV, IX, IX u. XIV). These modifications were fairly small, involving a terrace wall between the latrine and the *piscina* of the *frigidarium* and a different heating for the *tepidarium*, but leaving the latrines untouched. The latrines can be identified in rooms (XVI) and (XIX), both of which had a channel running through them and are at a distance to the bath proper. The entrance to the southern latrine (XIX) was through a corridor (XVIII) next to the *apodyterium* (XI). The northern latrine (XVI) had an entrance on the outside, ensuring that the bathers were not inconvenienced by the smell. Kretzschmer (1981: 622) allocated the southern latrine to the bathing guests and the northern to the staff.

In addition to studying the heating system of the bath, Fritz Kretzschmer (1981: 67–69) also studied the flushing system of the latrine and made a reconstruction (Figure 6b): Drain 152, which ran through the latrine, was build from rubble and covered with tiles. It started under the *frigidarium*, where it took up the spray water from the pool and the water from cleaning the room through a hole in the floor (see Piepers 1981: 37, fig. 11, pl. 22,1 u. 25,1). In addition to that, water from the overflow pool in the *tepidarium* could be introduced. The channel first flushed the southern latrine (XIX) and then ran on the western side of the bathhouse to the north, where it gathered the water from the *piscina* of the *frigidarium* and the *alveus* of the *caldarium*. The junctions were no longer preserved at the time of the excavations, but can be reconstructed with the help of the drains of the Zülpich bathhouse (Dodt 2003: 22–88, esp. 87–88; see above). The channel then ran through the northern latrine and finally towards the nearby Merzbach. The two latrines were thus connected via the channel system. Kretzschmer reconstructed the latrines after parallels in Vaison-la-Romaine and Ostia (Figure 6c). The channel in room XIX was L-shaped (Piepers 1981: Taf. 23,2). Because of the thickness of the preserved wall, Kretzschmer seems to interpret only the eastern part as having been used as a channel. But if we compare this to the L-shaped latrine of the *villa rustica* of Ahrweiler (no. 5), the southern part may also have had a seat bench, which would raise the number of possible seats from 3 to 5.

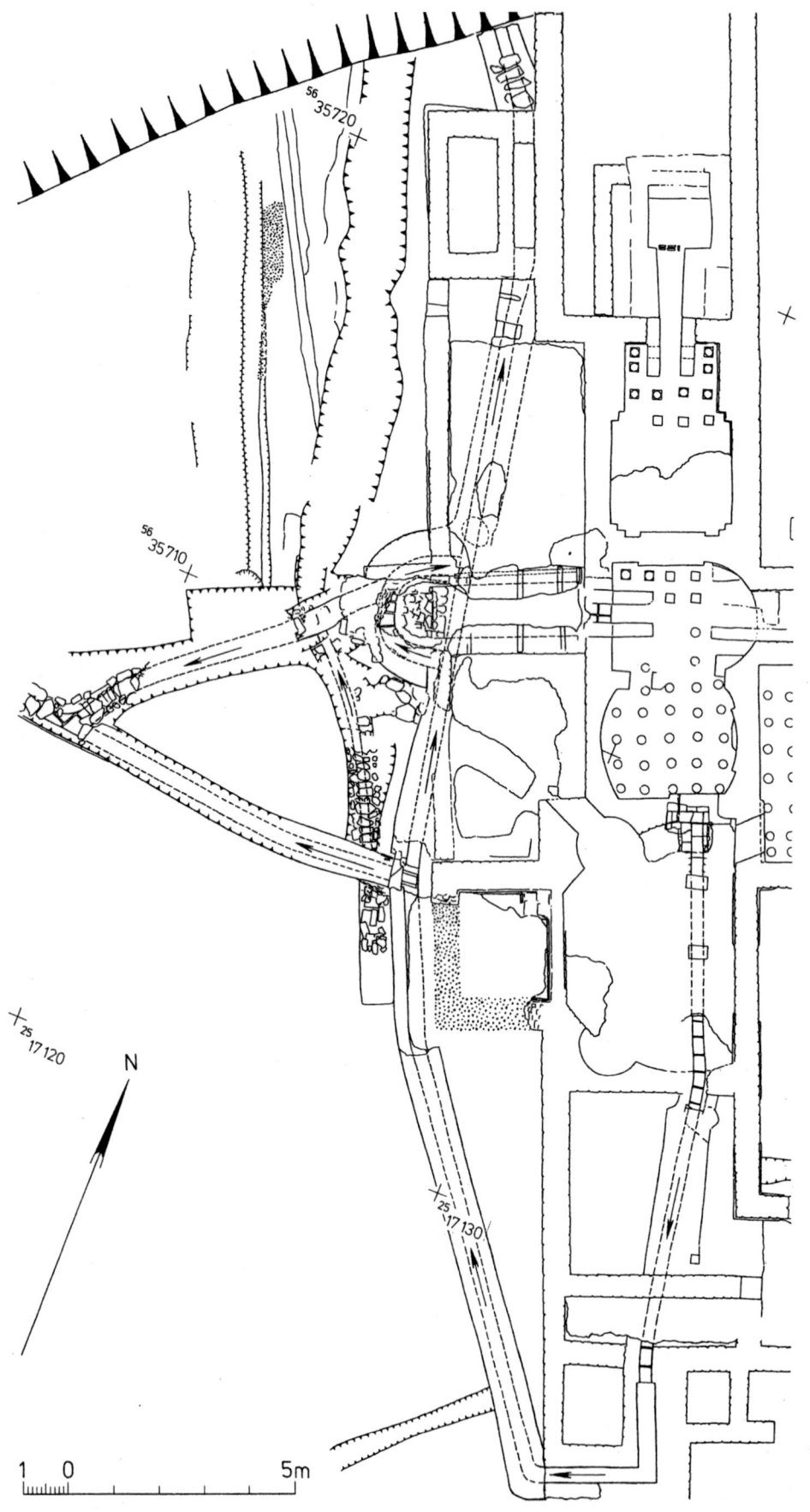

Figure 6a. Villa rustica of Lürken, excavation plan (Piepers 1981, Fig. 24. 26a. 29).

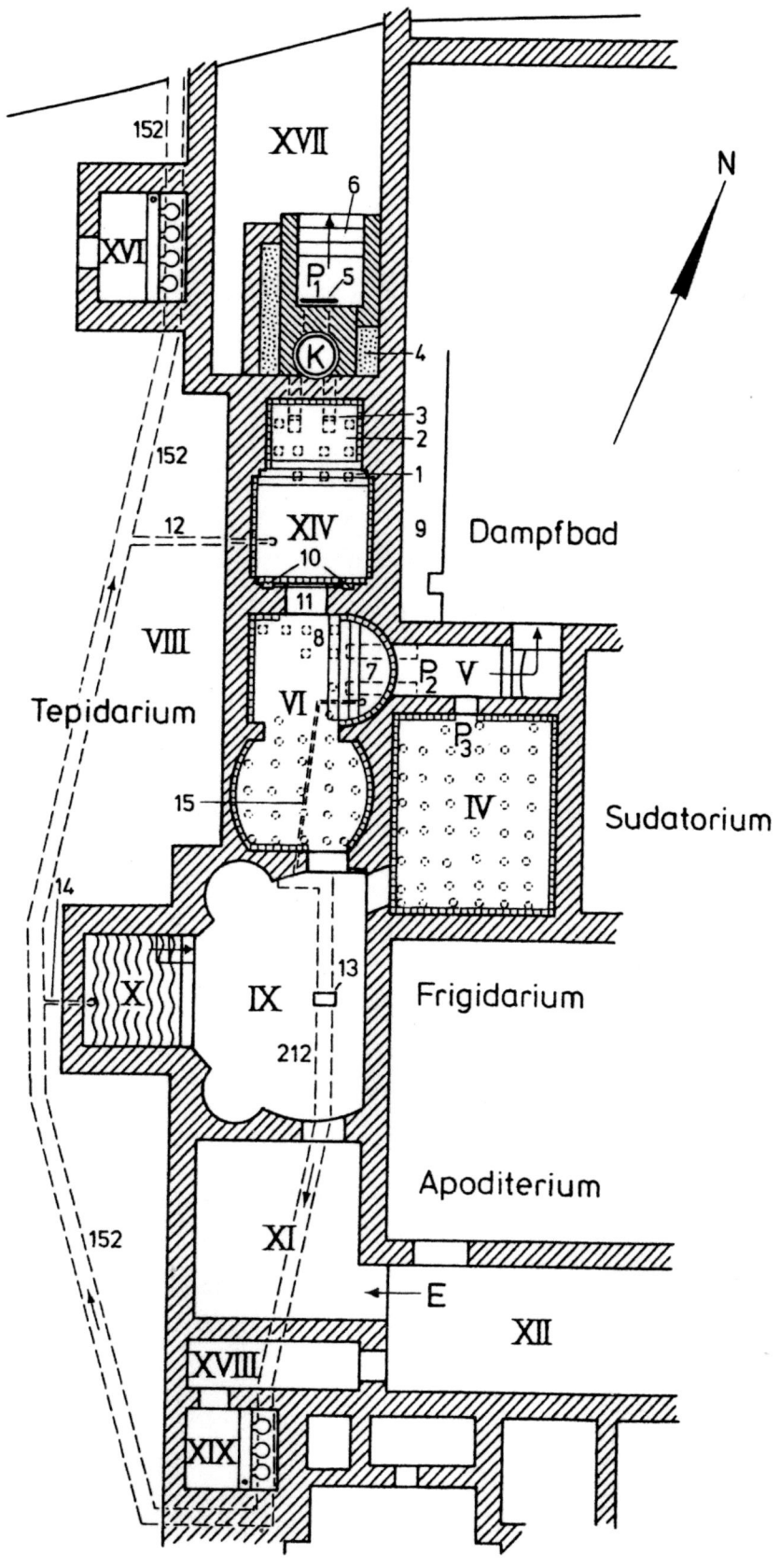

Figure 6b. Villa rustica of Lürken, reconstruction (Piepers 1981, Fig. 24. 26a. 29).

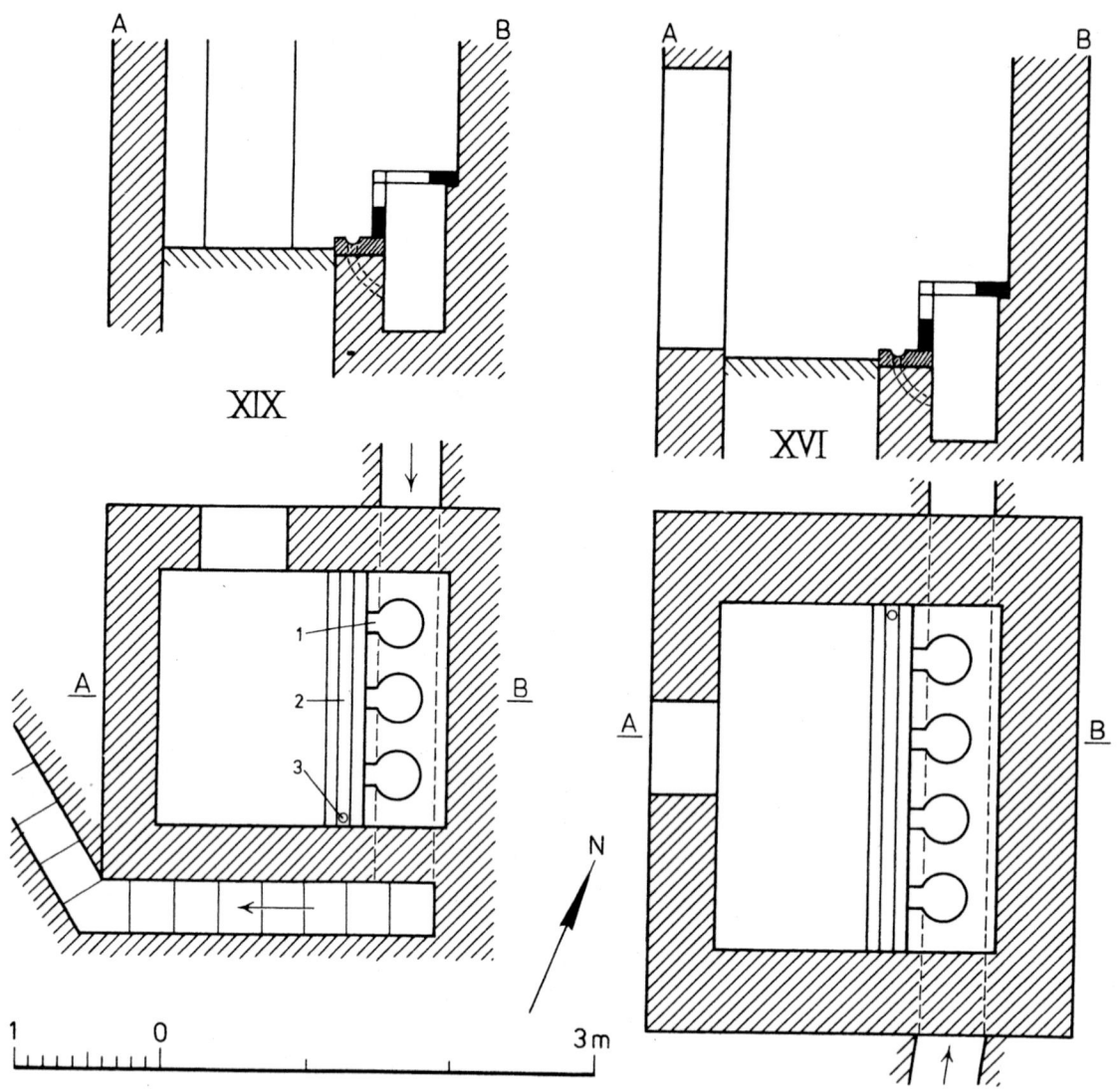

Figure 6c. Villa rustica of Lürken, rooms XVI and XIX (latrines) (Piepers 1981, Fig. 24. 26a. 29).

Latrine XVI on the outer wall of the bathhouse had an open drain along this wall (Piepers, footnote 30, pl. 23,1 and 25,1). The channel has a width of 0,45 m and is better preserved inside the room, because the little wall supporting the seat bench has been fortified, in contrast to the part of the channel outside of the building. Kretzschmer assumes 4 seats for latrine XVI and an entrance from the opposite side.

7) *Villa rustica* of Cologne-Vogelsang (Nordrhein-Westfalen/D)
(Figure 7a/b)

The *villa rustica* near Gut Vogelsang lay about 10 km west of Colonia Claudia Ara Agrippinensium/Köln. The main road between CCAA and Boulogne-sur-mer ran c. 3 km south of the villa. The bath was the main part excavated in 1989, as this had been well preserved due to the deep foundations of the *hypocaust* and the pools (Dodt 2001: 319–324; Dodt 2003: 220–223; Seiler 1993: 481–498). A part of the walls of the *frigidarium* with less deep foundations were completely destroyed (Figure 7a). The *piscina* (15) of the *frigidarium* had a floor only partly preserved and walls destroyed to almost floor level. This *piscina* is adjacent to a better-preserved floor (12), with a width of c. 1 m and a length of 1 m, which has a gradient of 20 cm to the south. Like the floor of the *piscina*, it is surrounded with walls on three sides and drains into a channel on the south side (24). The channel was only preserved as discoloration and could be traced for 1,50 m (Dodt 2003: 323 Figure 44) The better preservation of this context is proof that it is lower than the floor of the *piscina*. Because the *piscina* (15) lies slightly higher than floor (12), it has been destroyed to a higher degree, so that no drain or channel could be found. However, a drain used for the flushing of the room with floor (12) seems likely (Seiler 1993: 485-86). The source of the water may have been the well discovered during the excavations (Seiler 1993: 491; 498 Nr. 26). The trapezoid shape of

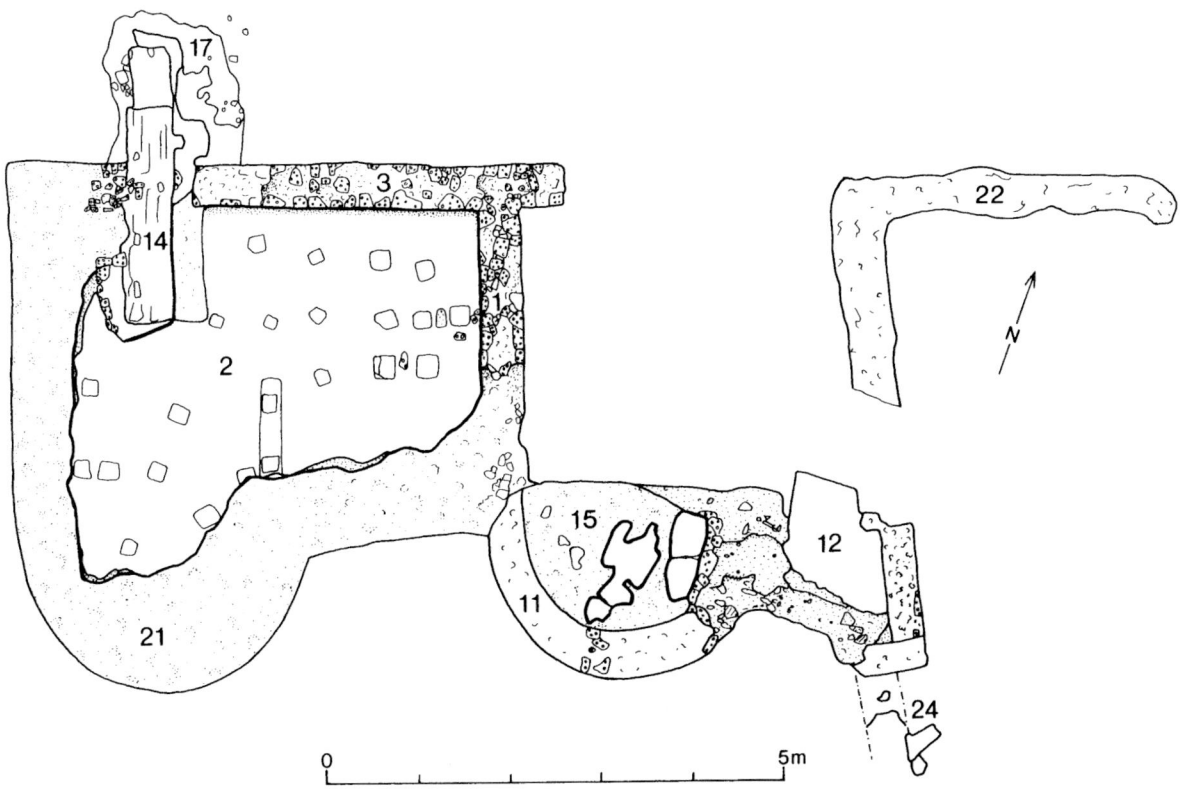

Figure 7a. Bath of the Villa rustica of Cologne-Vogelsang, excavation plan
(S. Haase, Römisch-Germanisches Museum Köln, after Dodt).

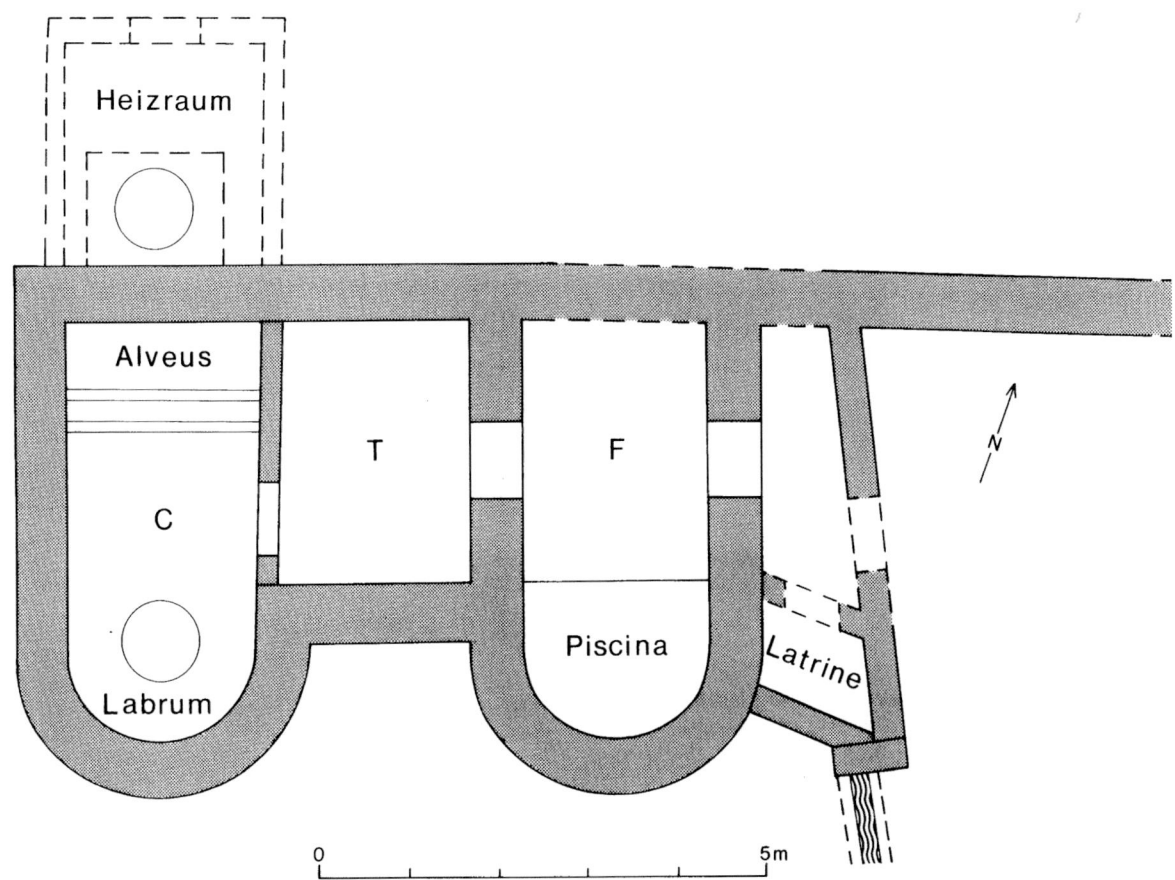

Figure 7b. Bath of the Villa rustica of Cologne-Vogelsang, reconstruction (S. Haase, Römisch-Germanisches Museum Köln).

the floor and its building method do not comply with the usual arrangements of Roman channel latrines, but the interpretation corresponds with the function of the room (Figure 7b). The eastern wall of the latrine has a breach, but continues as the angular foundation trench (22) (Figure 7a), whose EW part is in the same axis as the northern wall of the *hypocaust* room. On the NE edge of the floor of the latrine, a thin wall must have been erected. The latrine must have had an entrance through a door in this wall, from the narrow and tapering room to the north, interpreted as *apodyterium*. The entrance to the bath was probably in the eastern wall of the *apodyterium*, the door being placed where the breach in the wall is today, which would explain the placement of the northern wall of the latrine and the latrine's unusual shape.

8) *Villa rustica* of Blankenheim-Hülchrath (Nordrhein-Westfalen/D)
(Figure 8a/b)

The bath in the NW wing of the villa rustica of Blanckenheim, was very well preserved at excavation in 1894 and 1914 (Oelmann 1916: 212-226) A reconstruction of the building with a new definition of building phases was undertaken by Mylius (1933: 11-21) The whole complex has lost a lot of substance, due to having lain open after the excavation for some time (Jenter 2007: 137-139; Kunow 2011: 481-486). The walls of the villa were preserved up to 1 m above floor level. In the bath, some installations like the *suspensurae* of the heated rooms and the parapet of the pools were still preserved, giving us an idea about the real size of the rooms with wall heating and a good picture of the access to the different rooms, which again helped in visualizing the route the bathers must have taken (Dodt 2003: 262-267). During the second building phase of the main building, which can be dated to the 3rd and 4th century, latrine (65) was integrated into the bath block, but separated from the other bathing rooms – it is only connected to the *apodyterium* (52/58) by corridor (59/66) (Figure 8a), according to Mylius (1933: 11-21) to avoid unpleasant smells. Oelmann (1916: 220-225) assumes a third building phase here. A second entrance was formed by a little door leading outside and located in the wall across from the corridor. The length of latrine 65 corresponds to the length of *piscina* 68 in the neighbouring *frigidarium*. The latrine was also well preserved during excavation, which allowed a good documentation of the details of its installations (Figure 8b, Oelmann 1916: 225). The walls of the room were erected as double walls of small limestone ashlars filled with rubble; the latrine channel itself was made from plastered brick walls. This channel is between 1,00-1,25

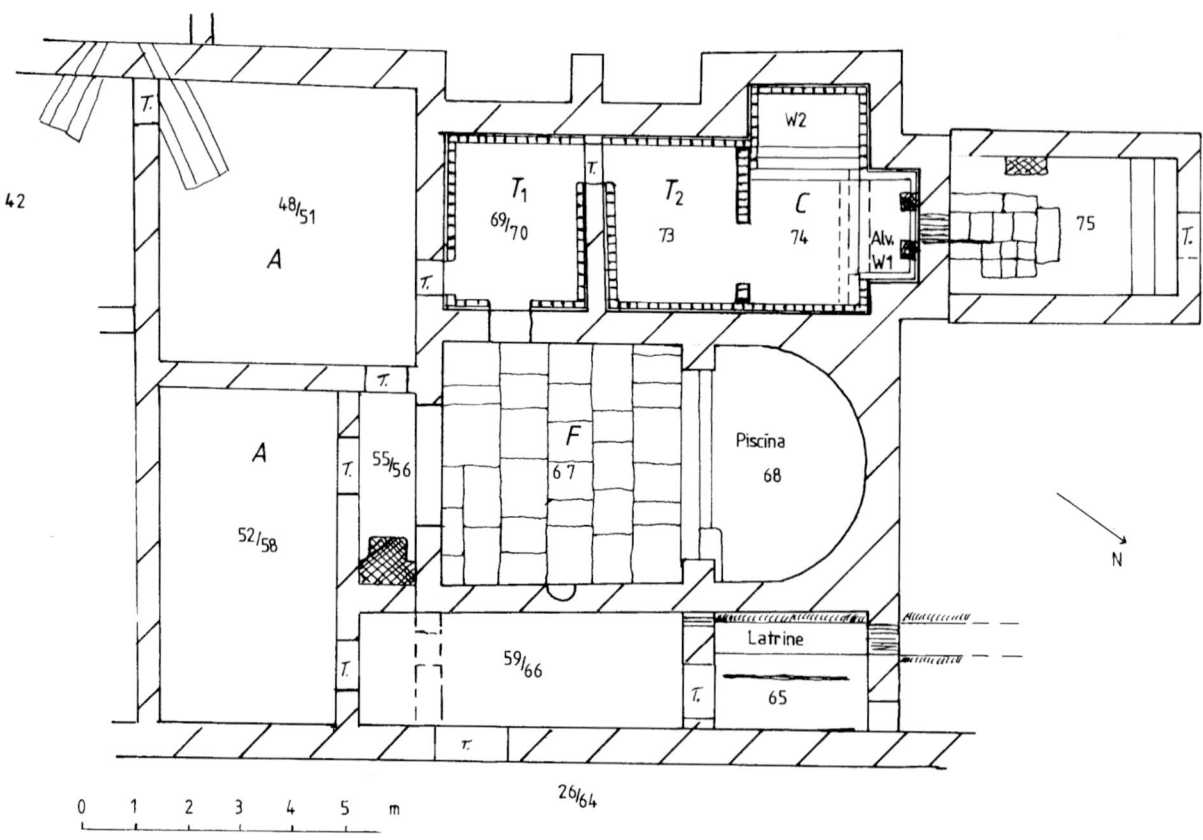

Figure 8a. Bath of the Villa rustica of Blankenheim, period IIb, plan of the excavated walls (Dodt after Oelmann 1916, Taf. 14).

Figure 8b. Latrine 65 at he bath of the Villa rustica of Blankenheim, seen from the north (after Oelmann 1916, Taf. 19,1).

m deep and 0,80 m wide and set against the long side of the latrine abutting the *frigidarium*. It begins in this room and has a gradient of 25 cm to the NW, where it flows into a channel on the outside of the building. The floor of the *piscina* sloped towards the NE and the wastewater drained through a lead pipe from the eastern corner into the channel in room (65) to flush the latrine. The room had a length of 2,90 m and a width of 2,15 m. If we compare this length with the length of 2,15 m of the latrine of the villa at Lürken (no. 5), for which Kretzschmer assumed 4 seats, we may safely assume at least 5 seats in latrine (65) of the *villa rustica* of Blankenheim.

Herman Mylius (1933: 15) suggested that during earlier building phases, the latrine may have been in the SE, under the corridor (59/66), but due to the widespread destruction of the earlier building phases, his reconstruction and interpretation cannot be verified.

9) *Villa rustica von Voerendaal* (Prov. Limburg/NL)
(Figure 9a/b)

Similar to the *villa rustica* at Ahrweiler (no. 5), the bathing complex of the *villa rustica* of Voerendaal has several building phases, which were discovered during the extensive excavation of the main building between 1947-1950, led by W. C. Braat (1953: 48-79). The bath had been erected at the beginning of the 2nd century and flanked the left side of the courtyard in front of the main building (Braat 1953: 64, Figure 11). The renovation cannot be dated precisely, nor is it certain if the different building activities belong to a single building period.

While the main building seems to have been provided with water by a well, the bathing complex seems to have received water from a channel diverting water from the nearby Hoensbeek, maintaining a constant flow (Dodt 2003: 238-244; Willems 1987: 49). Further to the south, on a lower position, the same creek received the wastewater from *piscina* (7) and the pool in the apse of *caldarium* (8) via masonry channels (Braat 1953: 61, pl. IX, 1 and 4). These channels were directly attached to the outer walls of the bath complex, with no room in between that might have been interpreted as a latrine.

During the second building phase, pool (2a) acquired an extension of channel β in the form of an open drain, tiled with *tegulae*. Another supply into the drain (α started at the NW corner of room (4/5), which had been added to the northern side of room (3). Due to a sloping drain fashioned from an *imbrex* in the NW corner of room (3), Braat (1953: 60; 62) assumed that a pool was

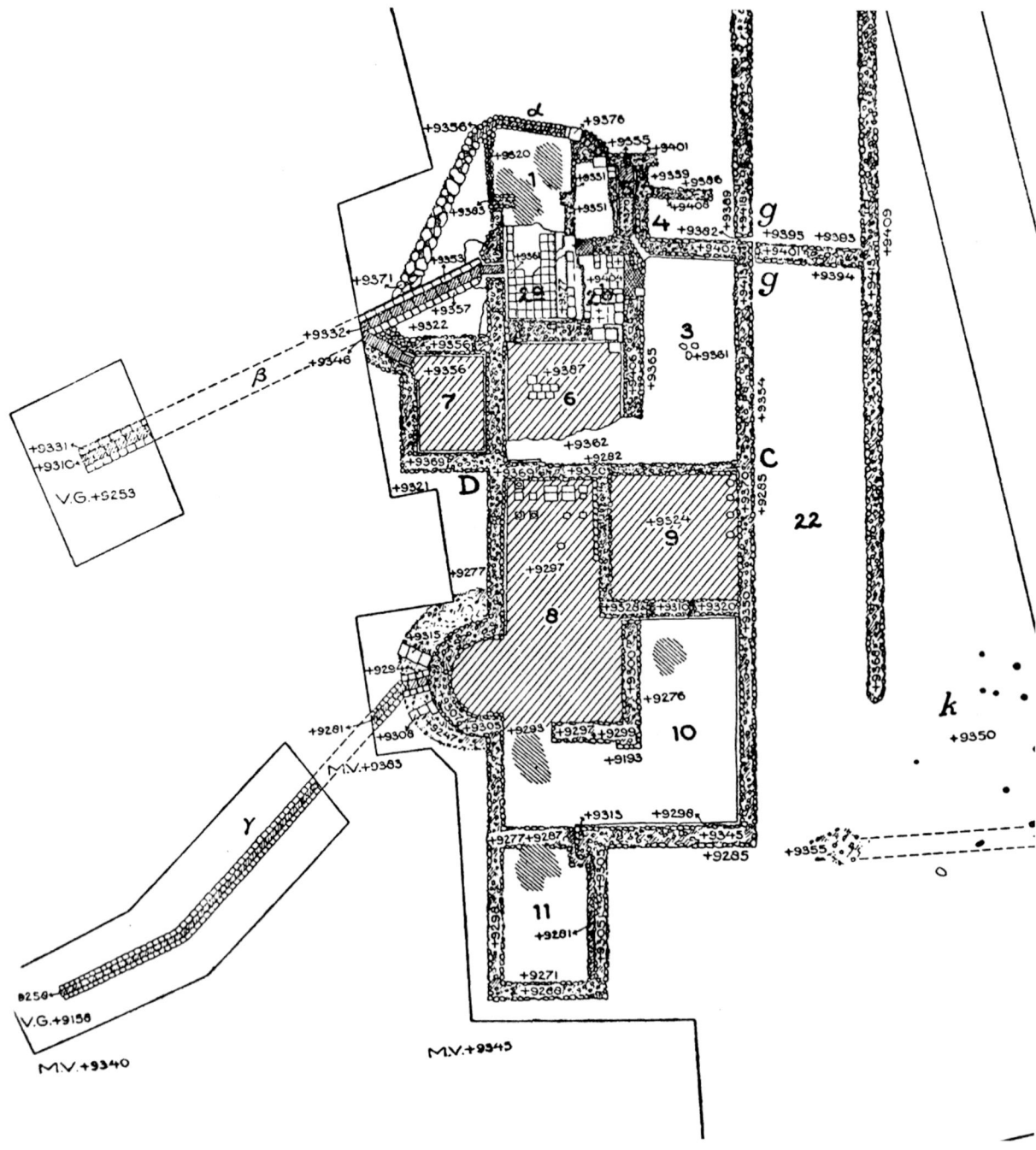

Figure 9a. Bath of the Villa rustica of Voerendaal, excavation plan (Braat 1953: 49 and. Taf. 8).

added to the northern side of this room. This room is interpreted by Braat as additional *frigidarium*. While the walls of most rooms and pools have been preserved, according to the published documentation, they seem absent in room (3). The pool was probably on the same level as the floor and approached over a parapet. The drain led to the lower open drain (5) on the west side of the room (Figure 9b). Channel (α was connected to the northern side of drain (5), running in NW direction. The construction of drain (5) conforms to a Roman latrine above an open drain with running water. Drain (5) connects to room (4), which is separated by a WE wall (Figures 9a and 9b). Perhaps it separated a southern part with two seats from a northern part with one seat, thus creating a double latrine.

In a later building phase, the connecting pipe was closed and the pool and latrine must have been abandoned.

Figure 9b. Bath of the Villa rustica of Voerendaal, seen from the west (Braat 1953: 49 and. Taf. 8).

10) *Villa rustica of Stolberg* (Nordrhein-Westfalen/D) (Figure 10)

The *villa*, in whose western tract a bathing complex was situated, was discovered in 1876, and excavated (after being looted in the years 1880/81) under the direction of the conservator of the Aachener Museums-Verein, F. Berndt (Berndt 1882: 179–188; Dodt 2003: 231–234; Reutti 1975: 217–221). Part of the villa was destroyed during the construction of Stolberg train station in the year 1888.

The villa was probably erected at the beginning of the 2nd century, but latrine q in room Q belongs to a modification phase, as F. Berndt found out: Room Q and the circular room R had been set against room C (Berndt 1882: 182; Reutti 1975: 220). The bath of the main building was probably supplied with water from a spring in the slope (Reutti 1975: 219-20). The find of a lead pipe with a tap in the northern wall of room H demonstrates that the bath did have running water (Berndt 1882: 185). The only channel in the bath was discovered connected to the circular room R (diameter 1,60 m), hence this must have contained a pool. This arrangement is similar to one of the circular rooms of the villa bath of Bonn-Friesdorf (Dodt 2003: 255–258). Room R, a later addition to the bath, lay SW of the *frigidarium* (rooms N, NI und NII) and could be approached by corridor Y. South of room R was the contemporary room Q, which may have been approached by corridor Y or by room C. The latter belongs to the original building, while room Q was added later. From the circular room R, the wastewater was led through the channel in westerly direction. A branch (q) ran from the western wall of room Q southwards. From there, the wastewater could be led southwards down the slope to the Inde creek. Channel q was build from bricks and thus conforms to the Roman channel latrine model. But the size and form of room Q, which is almost square with a length of 4,50 m – and thus bigger than any of the other rooms of the bathing complex – gives rise to doubt the interpretation as a latrine. Channel q has double the length of the larger latrine of the *villa rustica* of Lürken (no. 6), allowing for 8 seats above the channel. This seems oversized for the small *villa rustica* of Stolberg (a comparison of *villa* sizes in *Germania inferior* can be found in Dodt 2003: 126-27). Perhaps room Q had a partition with a framework construction, which had not been recognized during the excavation more than 125 years ago. Because the

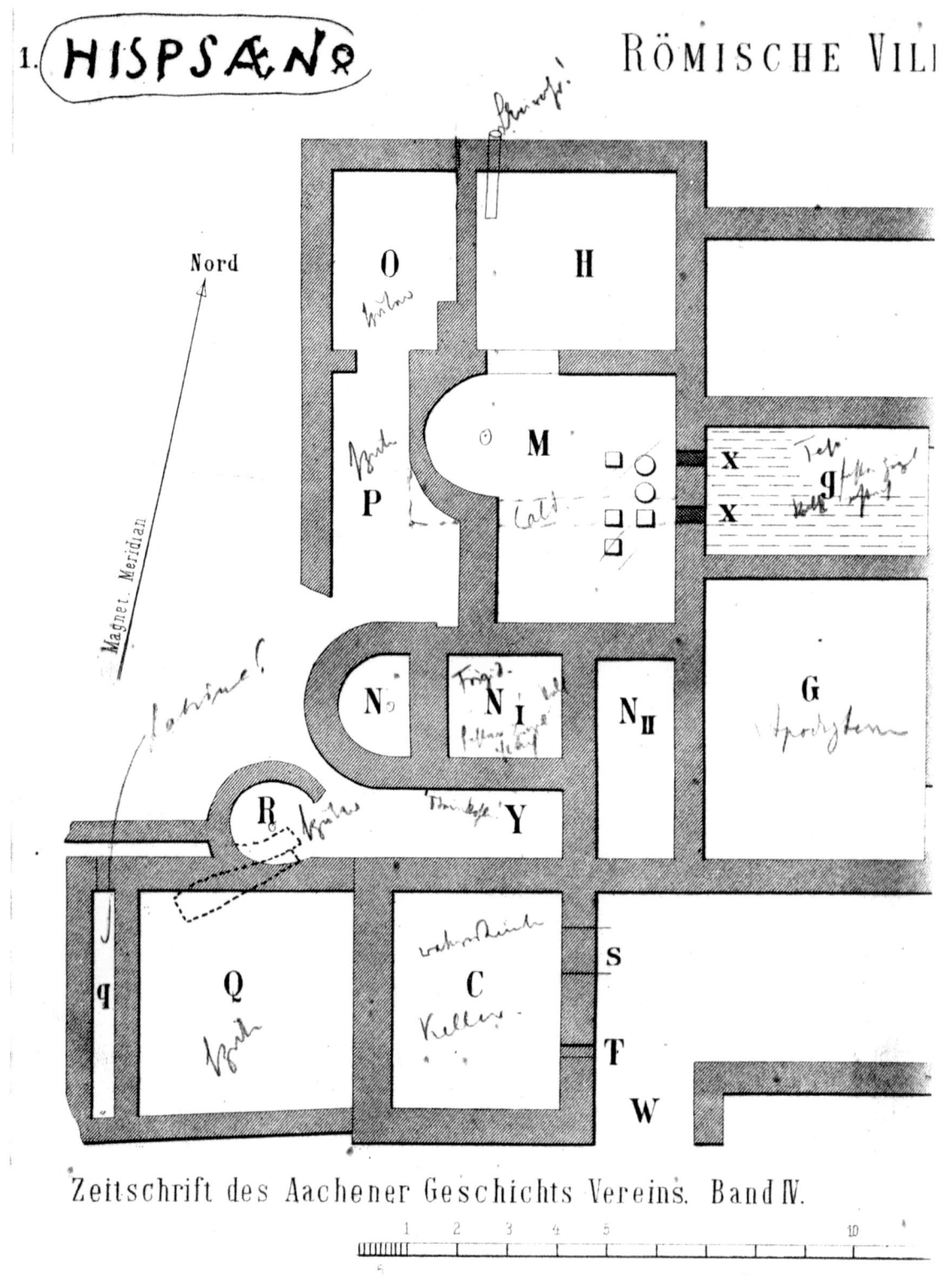

Figure 10. Bath of the Villa rustica of Stolberg, plan of the excavated walls (Berndt 1882: 188, Taf.).

rooms Q and R are later additions, the question of the draining of the wastewater from the pools in the *caldarium* and *frigidarium* and thus the question of the position of the latrine during earlier building phases has to remain unresolved.

11) ***Villa rustica* of Kreuzweingarten** (Nordrhein-Westfalen/D)
(Figure 11)

Like the Stolberg *villa*, the bathing complex of the *villa rustica* of Kreuzweingarten had a large room with a channel. The first excavation of this large and richly

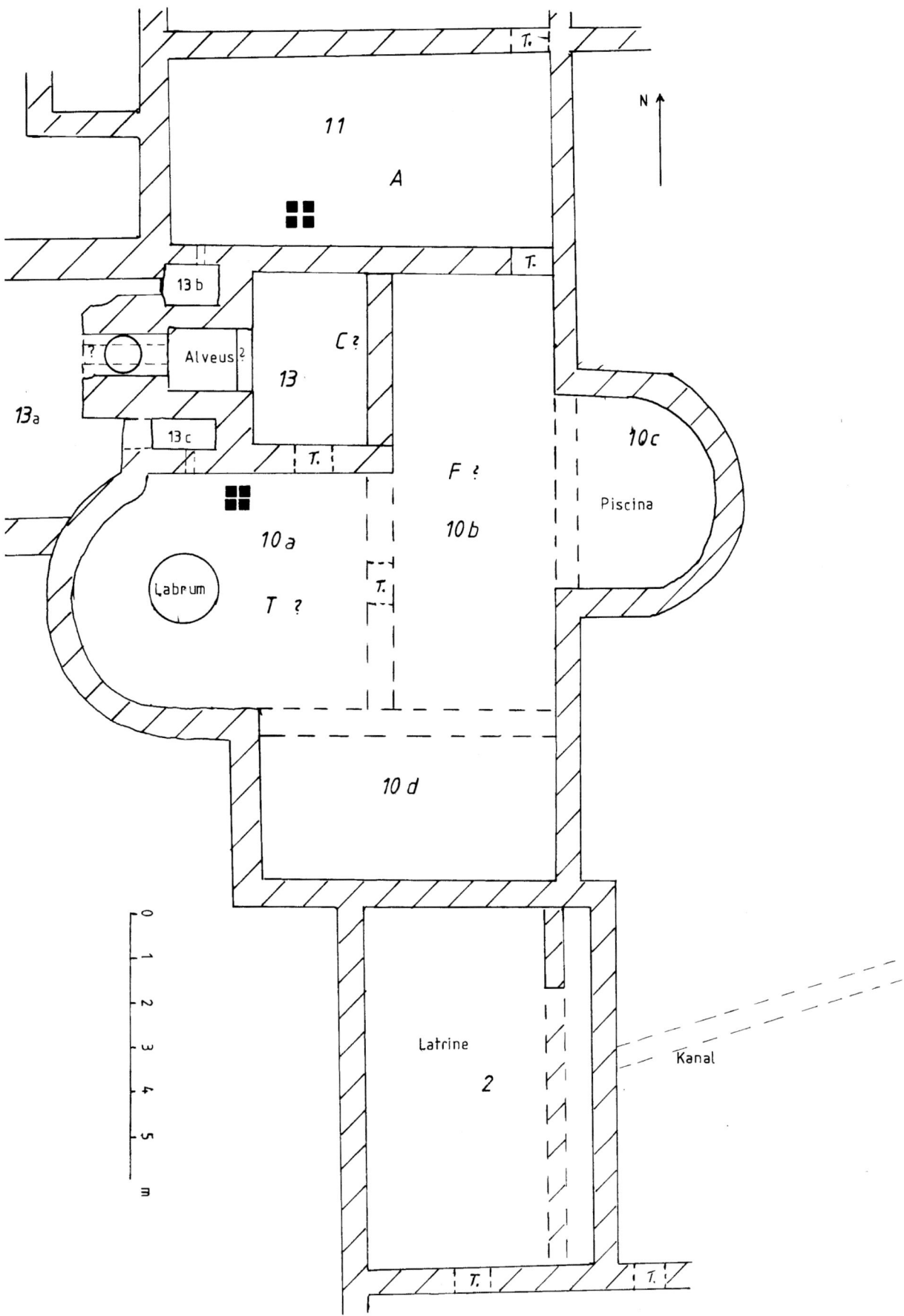

Figure 11. Bath of the Villa rustica of Kreuzweingarten, plan of the excavated walls (Dodt after aus'm Weerth 1900, Fig. 82).

appointed *villa* is even older than that of the Stolbeg *villa* (no. 10). The second excavation, which resulted in the addition of several rooms on the eastern and southern sides of the *villa* to the ground plan, is contemporary to the one at Stolberg (aus'm Weerth 1900: 187-191.; Dodt 2003: 270–273; Kunow 1987: 426-427.; Overbeck 1851; Piepers 1974: 152–158; Reutti 1975: 585-588). One of the newly excavated rooms was room 2, which may be interpreted as latrine. The main building had been erected in the first half of the 2nd century and was 60 m long, with two protruding tower-like structures (in German: *Risalit*) on the corner. The courtyard on the southern side with its surrounding tracts – among which was the bathing complex with the latrine – had been added later. The coins found date until the age of Constantine. The flagstones covering the channels in the southern part of the *villa rustica* had been made from calcareous sinter from the nearby *aqueduct* to Cologne (Grewe 1986: 114. 272-73). The bath (10) had not been documented in detail, complicating its interpretation. The interpretation of room (2) of the bath is equally uncertain, as its function was concluded from its location and the channels connecting it.

The villa was supplied with water from the northern slope by a stone channel, ensuring that both the bath and the latrine enjoyed a constant supply of running water. The small room (13) with its adjoining boiler room is only room of the bath interpreted with any certainty as *caldarium*. The large parts to the east and south of that room (10) were not subdivided or classified during the excavation (Koethe 1940: 130). It has to be taken into account that room (2) might have had a subdivision, which was not recognized during the excavation. As a part of room (10) seems to have a *hypocaust*, a subdivision into a *tepidarium* and a *frigidarium* with several pools or basins seems likely (Figure 11). It is conceivable that a pool in the southern part of room (10) drained its wastewater into the latrine in the southern adjoining room (2). The latter has a wall running parallel to the eastern wall of the room and a connection to a channel running from the middle of the eastern wall through the courtyard towards the Mühlbach creek on the eastern side of the villa. It is possible that the parallel walls formed a channel for the latrine, but room (2) has a total length of c. 8 m, which seems excessive for the private latrine of a family.

Summary and analysis

Currently, 14 latrines adjoining baths are known from *Germaina Inferior* (minus the Tungri region). Three of them have been recorded by Bouet in his handbook-cum-catalogue (see above), a further three are uncertain. Four of the 11 latrines presented in this little overview belong to public bathhouses, and seven to private baths in *villae rusticae*. Not a single channel latrine is known from the baths of the Roman camps and forts of the Lower Rhine limes, although, as in public baths, latrines must have belonged to the essential installations of these baths. The examples of Oberaden and Nijmegen show that latrines in military camps do not have to be connected to baths (Bouet 2009: 373-383).

We can thus summarize that out of the 13 public bathhouses (including mineral baths) documented for *Germania Inferior*, six have had latrines identified. In contrast, only six of the 74 documented *villa rustica* baths have latrines identified with confidence. The percentage of latrines in baths is thus higher in public baths than in private ones. It is plausible that channel latrines were present in all public baths, while *villae rusticae* might often have had only simple cesspit-latrines. The limitations of old excavations or the excavation being restricted to a section often stand in the way of a definite interpretation of a room as latrine. An example is a bath discovered in a relatively recent excavation in Köln-Nippes, which had been identified by the *alveus* and boiler room alone (Seiler 2006: 45–49 Figures 11–13). Consequently, the amount of data for latrines in *Germaina Inferior* is too small to allow far-reaching conclusions using statistics, as they might speedily be proven wrong by new excavations or research.

All of the latrines presented here have flushing channels. Most of these are fed by the wastewater from the pools, especially from the cold-water pools (*piscinae*) in the *frigidaria*, which are nearest to the latrines and often provide permanent flushing water. In order to avoid smells, the latrines are not directly accessible from the *frigidaria*. Latrine and *frigidarium* often share access through a corridor or the *apodyterium*. If latrines are connected to public bathhouses, it does not automatically follow that only the (paying) bathhouse visitors were able or allowed to use the latrine. The connection of bath and latrine was strictly practical and connected to the large amounts of wastewater accumulating in the baths, which could be put to use flushing the latrines. But other supplies were also possible; in the public bathhouse of Zülpich (no. 2, second building phase) and the bath of the *villa rustica* of Ahrweiler (no. 5), the latrines were flushed with freshwater: in Zülpich from a cistern fed by rainwater and in Ahrweiler from a drainage channel.

The latrine of Aachen (no. 4) is the only one allowing the reconstruction of the seat bench, as the outer walls of the building were preserved up to c. 1 m above the floor level. Most other latrines were identified as a room with a channel running through it, which was fed from a drain beginning at the pools of the bath. The seats of all the latrines must have been constructed above the channels like in Aachen (no. 4). The latrine at the bath of Cologne-Vogelsang (no. 7) is the only one with a small basin, the interpretation as latrine is based on the flushing system and the drain. The interpretation is thus relying on the water technique, even though the latrine does not conform to the usual building type of a channel-latrine.

Because of the small data set, no attempt will be made to relate the sizes of the baths and the latrines to each other, as too many different aspects would have to be factored in for a result. Up to now even the relationship between the size of a *villa rustica* and the size of its bath could not be ascertained (Dodt, 2006: 73–75). But it may be noted that the latrine of Aachen (no. 4) is remarkable for its number of seats (between 40 and 65) and the bath at Lürken (no. 6) is noteworthy for the fact that it has two latrines on a single channel. Perhaps the channel of the latrine in Voerendaal (no. 9) was also separated into two sections with three seats in all. The latrine in Aachen includes the technical installation of a sluice gates for flushing, a detail that is also found in one of the two latrines in the public baths at Xanten. Another significant observation is the fact that most of the latrines did not belong to the first building phase of the baths, but are later additions.

(Translated by Stefanie Hoss)

Bibliography

aus'm Weerth, E. 1900. Kreuzweingarten, römische Anlage und Funde. In P. Clemen (ed.), *Die Kunstdenkmäler der Rheinprovinz IV, 2. Die Kunstdenkmäler des Kreises Euskirchen*: 187–191. Düsseldorf.

Bechert, T. 1982. *Römisches Germanien zwischen Rhein und Maas. Die Provinz Germania inferior*. Zürich, München.

Berndt, F. 1882. Eine römische Villa bei Stolberg. *Zeitschrift Aachener Geschichtsverein* 4: 179–188.

Bouet, A. 2003. *Thermae Gallicae. Les thermes de Barzan (Charente-Maritime) es les thermes des provinces gauloises* (Aquitania Supplement 11). Bordeaux.

Bouet, A. 2009. *Les latrines dans les provinces gauloises, germaniques et alpines* (Gallia Supplement 59). Paris.

Braat, W. C. 1953. De grote Romeinse villa van Voerendaal. *Oudheidkundige Mededelingen* 34: 48–7

Cüppers, H. 1982. Beiträge zur Geschichte des römischen Kur- und Badeortes Aachen. In H. Cüppers, W. Sage und L. Hugot, *Aquae Granni. Beiträge zur Archäologie von Aachen* (Rheinischen Ausgrabungen 22) Köln.

Dodt, M. 2001. Römische Badeanlagen in Köln. *Kölner Jahrbücher* 34: 319–324.

Dodt, M. 2003. *Die Thermen von Zülpich und die römischen Badeanlagen der Provinz Germania inferior* (online publication of PhD thesis University of Bonn see http://hss.ulb.uni-bonn.de/2003/0117/0117.pdf).

Dodt, M. 2006. Bäder römischer Villen in Niedergermanien. Ausgrabungen im rheinischen Braunkohlerevier. *Bonner Jahrbücher* 206: 73–75.

Dodt, M. 2007. Römische Badeanlagen in Niedergermanien - eine Verbreitungskarte zum aktuellen Forschungsstand. *Archäologie im Rheinland* 2006: 96–99.

Dodt, M., 2008. Die Bauperioden der Zülpicher Thermen. *Archäologie im Rheinland* 2007: 107–109

Fehr, H. 1993. *Römervilla. Führer durch die Ausgrabungen am Silberberg Bad Neuenahr-Ahrweiler* (Archäologie an Mittelrhein und Mosel 7) Koblenz.

Garbrecht, G., Manderscheid, H. 1994. *Die Wasserbewirtschaftung römischer Thermen. Archäologische und hydrotechnische Untersuchungen* (Mittelungen des Leichtweiss-Institut 118). Braunschweig.

Grewe, K. 1986. Atlas der römischen Wasserleitung nach Köln, *Rheinischen Ausgrabungen* 26, 114: 272–273.

Haupt, D. 1984. Ein römischer Töpferbezirk bei Soller. (Rheinische Ausgrabungen 23) Köln

Heinz, W. 1979. *Römische Bäder in Baden-Württemberg. Typologische Untersuchungen*. Tübingen.

Heinz, W. H. 1983. *Die römischen Thermen*. München.

Hugot, L. 1959. Jahresbericht 1956–1968. *Bonner Jahrbuch* 159: 276–280.

Jamar, J. T. J. 1981. *Heerlen, de Romeins thermen*, Zutphen.

Jamar, J. T. J. 1988. *Römisches Leben in Heerlen. Ausstellungskatalog Mönchengladbach – Venlo – Dormagen – Maaseik*.

Jenter, S. 2007. Die Villa rustica in Blankenheim, *Archäologie im Rheinland* 2006: 137–13

Koethe, H. 1940. Bäder römischer Villen im Trierer Bezirk. *Berichte der Römisch-Germanischen Kommission* 30: 43–131.

Kretzschmer, F. 1981. Das Römerbad in Lürken aus technischer Sicht. In Piepers 1981: 51–73

Kunow, J. 1987. Euskirchen-Kreuzweingarten. Römischer Gutshof. In H. G. Horn (ed.), *Die Römer in Nordrhein-Westfalen*, Stuttgart, 426-427.

Kunow, J. 2011. Römische Badeanlagen im Rheinland. Aktuelle Planungen und Realisierungen zu ihrer Präsentation und Erschließung im städtebaulichen Kontext. In M. Müller, Th. Otten and U. Wulf-Rheidt (eds), *Schutzbauten und Rekonstruktionen in der Archäologie. Von der Ausgrabung zur Präsentation. Kolloquium Xanten 21.-23. Oktober 2009.* (Xantener Berichte 19): 481–486. Mainz.

Mylius, H. 1933. Zwei neue Formen römischer Gutshöfe (Villa von Blankenheim). *Bonner Jarhbücher* 138: 11–21.

Nielsen, I. 1993. *Thermae et Balnea. The Architecture and Cultural History of Roman Public Baths*[2]. Arhus.

Oelmann, F. 1916. Die römische Villa bei Blankenheim in der Eifel. *Bonner Jahrbücher* 123: 212–226.

Overbeck, J. 1851. *Die römische Villa bei Weingarten* (Winckelmanns-Programm des Vereins der Altertumsfreunde im Rheinlande 7). Bonn.

Piepers, W. 1974. Die römische Villa von Kreuzweingarten. In *Führer zu vor- und frühgeschichtlichen Denkmälern* 26: 152–158. Mainz.

Piepers, W. 1981. *Ausgrabungen an der alten Burg Lürken* (Rheinische Ausgrabungen 21). Köln, Bonn.

Reutti, F. 1975. Römische Villen in Deutschland, PhD dissertation, Univesity of Marburg /Lahn, published on Microfiche.

Seiler, S. 1993. Eine Villa rustica in Köln-Vogelsang. Ausgrabung in einem römischen Gutshof. *Kölner Jahrbücher* 26: 481–498.

Seiler, S. 2006. Das römische Gutshaus am Weiher. In R. Kruse (ed.), *Der Nippeser Weiher. Nippes – Bemerkenswertes und Unterhaltsames aus einem Kölner Stadtteil* 6: 45–49 fig. 11–13. Köln.

van Giffen, A.E., Glasbergen, W. 1948. Thermen en castella te Heerlen/Coriovallum. *L'Antiquité classique* 17: 199-262

Willems, W. J. H. 1987. De grote villa van Voerendaal. In *Langs de weg, Heerlen*. Maastricht.

Zieling, N. 2008. Die Thermen. In M. Müller, H.-J. Schalles and N. Zieling (eds), *Colonia Ulpia Traiana. Xanten und sein Umland in römischer Zeit. Geschichte der Stadt Xanten* (Xantener Berichte Sonderband 1): 373-38. Mainz.

Roman toilets in Nijmegen, *Oppidum Batavorum* and *Ulpia Noviomagus*, the Netherlands

Elly N.A. Heirbaut

1. Nijmegen: two civil centres throughout the Roman period

The first three centuries AD were an eventful period for Nijmegen. After the arrival of the Roman army, several forts and military camps were built and abandoned again in the eastern part of what is now the modern city, and on two locations along the river Waal, Roman towns grew up.

At first, the civil centre was in what is now still the centre of the city of Nijmegen. *Oppidum Batavorum* originated as a ribbon village, established around 10 BC to house the first veterans of the legions encamped in *Germania Inferior*. In the following decades, the village developed into the chief settlement of the *civitas Batavorum*. Besides veterans, Gallo-Roman civil servants, artisans, traders and shopkeepers also lived here. The central part of this settlement seems to have been fairly densely populated (Van Enckevort and Heirbaut 2010: 50-51). The most important central axis of this settlement was oriented east-west. South of this main road, in the St. Josephhof, several plots have been excavated suggesting that this part of the settlement was carefully planned and laid out. It seems probable that this was the case for large parts of the settlement. The actual buildings were situated directly on this main road; behind them, very long rectangular plots of land were used as gardens. In those garden areas, outbuildings and toilets (cesspit, sometimes superimposed by a structure), as well as postholes and ditches have been found.

Around 69/70 AD, *Oppidum Batavorum* was destroyed by a fire started by the Batavians, during the Batavian Rebellion. This did not, however, mean the definitive end of Roman presence at the site (Van Enckevort 2010: 241-252). Foundations have been found that had been dug through the burnt layer associated with the insurrection, which seem tot constitute a very fragmented ground plan of a large building. This was a so-called courtyard building, and has been tentatively identified as the residence of a commanding officer within the fortification, probably that of the Second Legion, present at Nijmegen in 70/71 AD (Van Enckevort 2010: 251-252).

After *Oppidum Batavorum* was destroyed by fire in 69/70 AD, the centre of Roman occupation shifted to the lower lying western part of Nijmegen. There a Roman town sprang up in the Flavian period that was to remain occupied until the 3rd century (*Ulpia Noviomagus*). During this period, the layout of the city underwent several changes, sometimes induced by external factors. Only the southern part of the city is well investigated, but unfortunately the only excavation that has been well published is the one in the south-western corner of the city, situated at the Rijnstraat and Lekstraat (Heirbaut 2013). Here, several plots have been excavated. The buildings were all orientated on the main north-south road. Behind them, long gardens stretched out over up to 120 m, housing small outbuildings, cesspits, pits and ditches.

Around 280 AD, a new *castellum* was built on the Valkhof hill. It seems that some decades before, around 240 AD, at least the southern part of *Ulpia Noviomagus* was largely deserted. As yet, little or no research has been conducted in the northern and central parts of *Ulpia Noviomagus*, and it is possible that these areas remained occupied after the southern zones had been abandoned.

2. Private lavatories: from indoor to outdoor

Up until now, research into toilets in *Oppidum Batavorum* (10 BC-69/70 AD) has been restricted to the excavation at the St. Josephhof (Figure 1, Heirbaut 2010). Excavations conducted prior to 2002 have not (yet) been published; excavations conducted in the period 2002-2010 were on too small a scale to yield usable information on toilets.

Several plots have been excavated here, oriented towards the main road to their north. On each plot there was a house, the layout of which conformed originally to the same pattern across all plots. A minimum of three building phases can be identified, but since the fronts of the houses were outside the excavation area it is difficult to identify the exact extent of each phase. The floor plans indicate that each house was divided into several plots positioned one behind the other, all having the same dimensions (approximately either five or seventeen feet). Between the third and the fourth and between the fourth and the fifth plot was a small corridor, oriented east-west. There were also hallways with a north-south orientation, which were somewhat wider. Although a minimum of three building phases have been identified, the first five plots seem to have been part of the original construction.

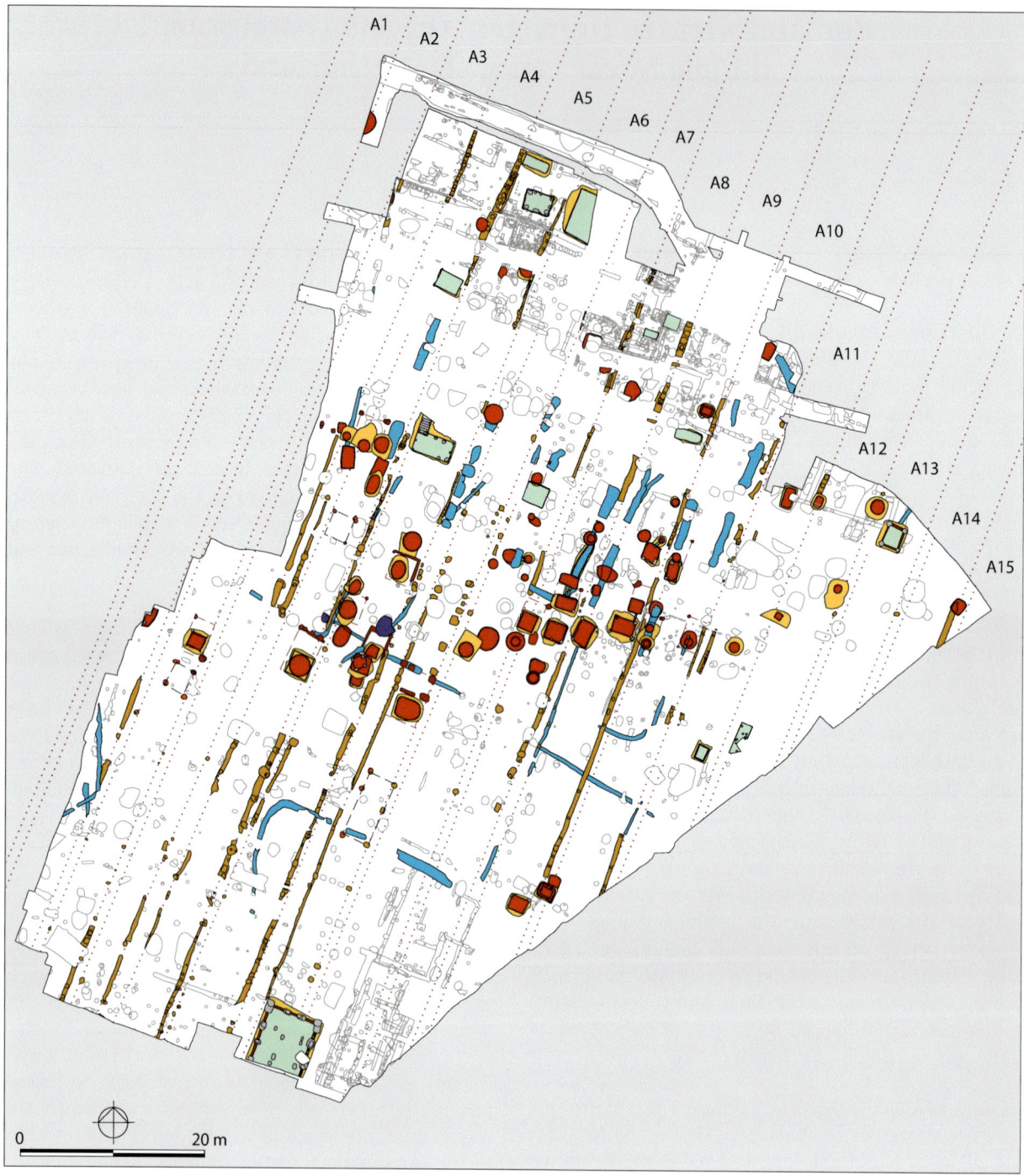

Figure 1. Feature map with cesspits, cesspools, overflow ducts and gutters for conducting water at St. Josepphof (*Oppidum Batavorum*) (©EH/TW)

On the basis of their location, two types of cesspits can be discerned: indoor and outdoor. Indoor cesspits have been found both in different plots and at different locations inside the plots (Figure 2). Cesspits have been found on the east and on the west side of the plot as well as positioned centrally in the house. In some cases the cesspit was on the border between plots and may have been used simultaneously by two households. Thus, on one plot no cesspit has been found, whereas one of the cesspits on the adjoining plot straddles the border between them, suggesting that this cesspit was used by both households. Probably, one of the plots had only a seating facility with a gutter connecting it to the cesspit. This is not an unknown phenomenon, at Pompeii, several households shared the same cesspit (Jansen 1994: 36).

Still, the fact that not a single cesspit has been found elsewhere in the excavated part of the house is remarkable; perhaps small barrels with a seat were

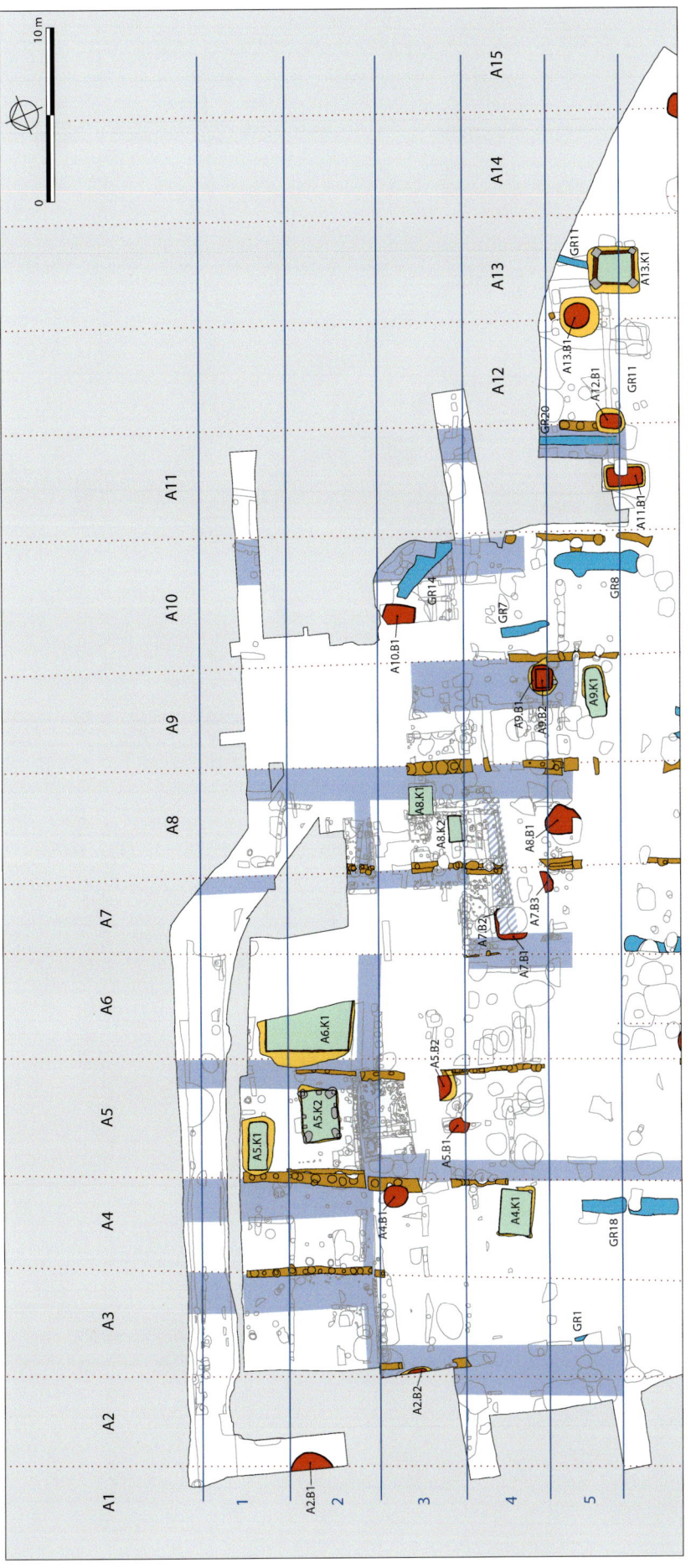

Figure 2. Northern part of the feature map, showing the different plots. property boundary (light brown), cesspits (reddish brown), gutters for conducting water (blue), corridors (light blue), cellars (green) (©EH/TW)

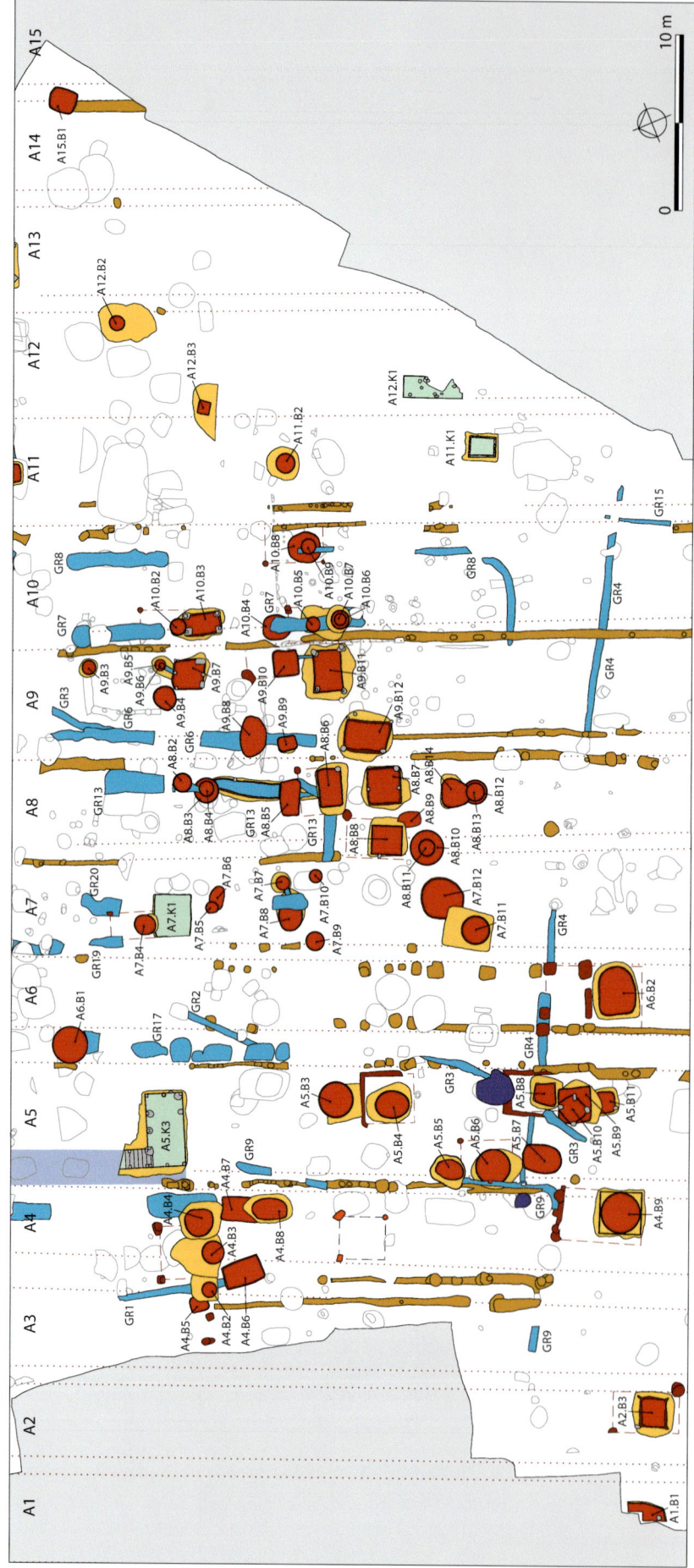

Figure 3. Middle part of the feature map. Property boundary / foundation ditch (light brown), cesspits (reddish brown), gutters for conducting water (blue), privies (reddish brown with dotted line), cellars (green), outbuildings (red with dotted line) (©EH/TW)

used, which were placed directly on the floor; these would have left no trace in the archaeological record. Also, chamber pots may have been used.

As far as latitude is concerned, it appears that not all plots had an equal number of cesspits. In some plots, not a single cesspit has been found in any of the houses; in other plots, cesspits were in some houses but not in all. On the basis of their location inside the house, their stratigraphical position in relation to other features constructed over them (such as foundation trenches), and to datable finds, these indoor cesspits all seem to belong to the Augustan-Tiberian period, in contrast to the outdoor cesspits, about which more will be said later. At first sight, the cesspits on plots A5 en A7 appear to be inside the house; however, on the basis of finds of window glass and mural decoration and the remains of a cellar built over them (all indications of either the back of the house or the existence of a room) both cesspits were outside, among the first to be so placed; probably in the early Claudian-Neronian period.

Although the plots are identical in shape, their use was evidently different according to the plot concerned. Indoor lavatories will never have been constructed in or near the most lavish rooms of the building, as evidenced by Cicero (De natura deorum II, lvi.— lvii): 'And just as architects regulate the drains of the houses to the rear, away from the eyes and nose of the masters, since otherwise they would inevitably be somewhat offensive, so nature has banished the corresponding organs of the body away from the neighbourhood of the senses'.

Rather, they will have been hidden in a dark corner of the house, such as beneath the stairs or at the end of a hallway, as at, for instance, Pompeii. Moreover, there are several examples of cesspits in kitchens. Again, Pompeii is comparable: Out of a total of 170 known toilets in Pompeii, 33 were in kitchens, 18 directly next to one, and 22 in gardens (Jansen 1994: 31, 36;7 Jansen 1997: 128)

However, the function and purpose of most of the rooms at the St. Josephhof remains unknown. In some cases the cesspit was constructed in a separate room dedicated to it exclusively: a toilet as we still know it today. Several cesspits were near or at the end of a corridor. In those instances where the cesspit actually is in the corridor, it would have blocked access if the seating had been directly above it. However, in the case of two cesspits on plot A9, a small room has been identified directly to their west. Possibly, seats had been placed there, with a small gutter conducting urine and faeces to the cesspit. The main advantage of a cesspit in the corridor or in a small room adjoining it would have been ease of access for cleaning and the removal of their rather smelly and not particularly pleasant contents without having to traverse other rooms. The same can be said about cesspit A12.B1, which is only partly indoors, as the back of the house was constructed across it. A similar phenomenon has been found at the *House of the Silver Wedding* at Pompeii: since the cesspit there was easily accessible from the garden, it was not necessary to enter the house each time the cesspit had to be emptied (G. Jansen *viva voce*).

Shortly before AD 40, the plots were shortened by the construction of an east-west alley to facilitate the development of new parcels south of it. Little is known about the buildings on these plots, however.

The transition from the Tiberian to the Claudian era marks the end of the custom of constructing indoor cesspits. Cesspits inside the houses were filled up, and from now on, cesspits were dug in the backyard area (Figure 3). Again, some plots did not have any cesspits at all. To some extent, this may be explained by the fact that the plots have only partly been excavated, but the question poses itself whether part of the explanation may lie in the fact that the building on the plot acquired a new or different function (i.e., no longer being used to live in)?

The spatial layout of the backyards differed sharply from parcel to parcel. On most plots, cesspits were near the border with its neighbours, but in some instances they were positioned more centrally on the plot. When installations were against the western border, in many instances they were in the 1.3 to 2.5m wide alleys, again facilitating the removal of their contents. The toilet seats, however, will not have been placed in the alleys, since that would have blocked access. More likely, the seats will have been placed on the east side of the garden wall, inside the garden. It is clear, however, that locations vary, without a clear pattern emerging; some were in alleys, some were at the front of the plot, some at the back and some in the centre. Moreover, the finds indicate that there was no clear pattern in the construction of these cesspits either; at any time, any cesspit could be constructed anywhere (Figure 4).

After the Batavian insurrection, the focal point of the civil habitation at Nijmegen shifted to the west, where from the Flavian period onwards the small settlement there developed into a fully-fledged Roman town: *Ulpia Noviomagus*. In the south-western corner of this town, a nearly complete *insula* has been excavated, on which several long rectangular parcels have been identified (Figure 5). The *insula* has been constructed gradually from AD 100 onwards, and was used for living quarters. Once again, the houses have only been partly excavated; the front parts lining the street remain largely unknown. Still, it has become clear that there were no indoor toilets; all cesspits were situated in the backyards. The emerging tendency to move toilets

Figure 4. View on some cesspits on plot A7 and A8. The surveying poles approximately indicate the boundary between the two plots. On the bottom, three cesspits in section (©BLAN/RM).

outside of the house thus continued, at least as far as could be established for this part of *Ulpia Noviomagus*. Each parcel had at least one cesspit, but most had more, one replacing the other as time progressed. Since dating material is lacking it is difficult to assign cesspits to specific periods, but it seems that the oldest cesspits are those close to the buildings, and that cesspits were progressively moved towards the rear of the plots as buildings were extended. Thus, the youngest cesspits are those closest to the rear of the plots. This can be deducted from the several building phases per parcel, with wall ditches and features belonging to the buildings cutting through cesspits. It suggests a more systematic approach, contrary to the pre-Flavian *Oppidum Batavorum*, where cesspits were dug across plots from front to back, back to front and left to right.

After this part of the town went up in flames at the end of the 2nd century, the area was parcelled out anew. The new plots all had a width of 6 m, and the buildings seem to have been standardized to some degree as well. Hardly any cesspits belonging to this period have been found, however, with the exception of two cesspits on parcel B4 and one on B3. Contrary to the situation in the previous period, cesspits were again sometimes placed inside the house. This is clear from cesspit B3.1, which was evidently dug in a room specifically designed for it.

3. Private lavatory infrastructure

3.1 Toilet buildings

To ensure some degree of privacy, each parcel will have had one or more toilet buildings (Figure 6). In *Oppidum Batavorum* these small square or rectangular wooden structures were observed in several places, as small postholes forming a square or rectangular outline, or as shallow wall ditches (Heirbaut 2010), both situated a short distance from a cesspit. Moreover, the abandonment of a cesspit did not necessarily imply the abandonment of the privy. The part of the lavatory above ground could continue to function, the connection to the new cesspit being made by way of a small gutter.

Apart from a privy on parcel A7, no archaeological remains of toilet buildings have been found in the south-western corner of *Ulpia Noviomagus*. Following the

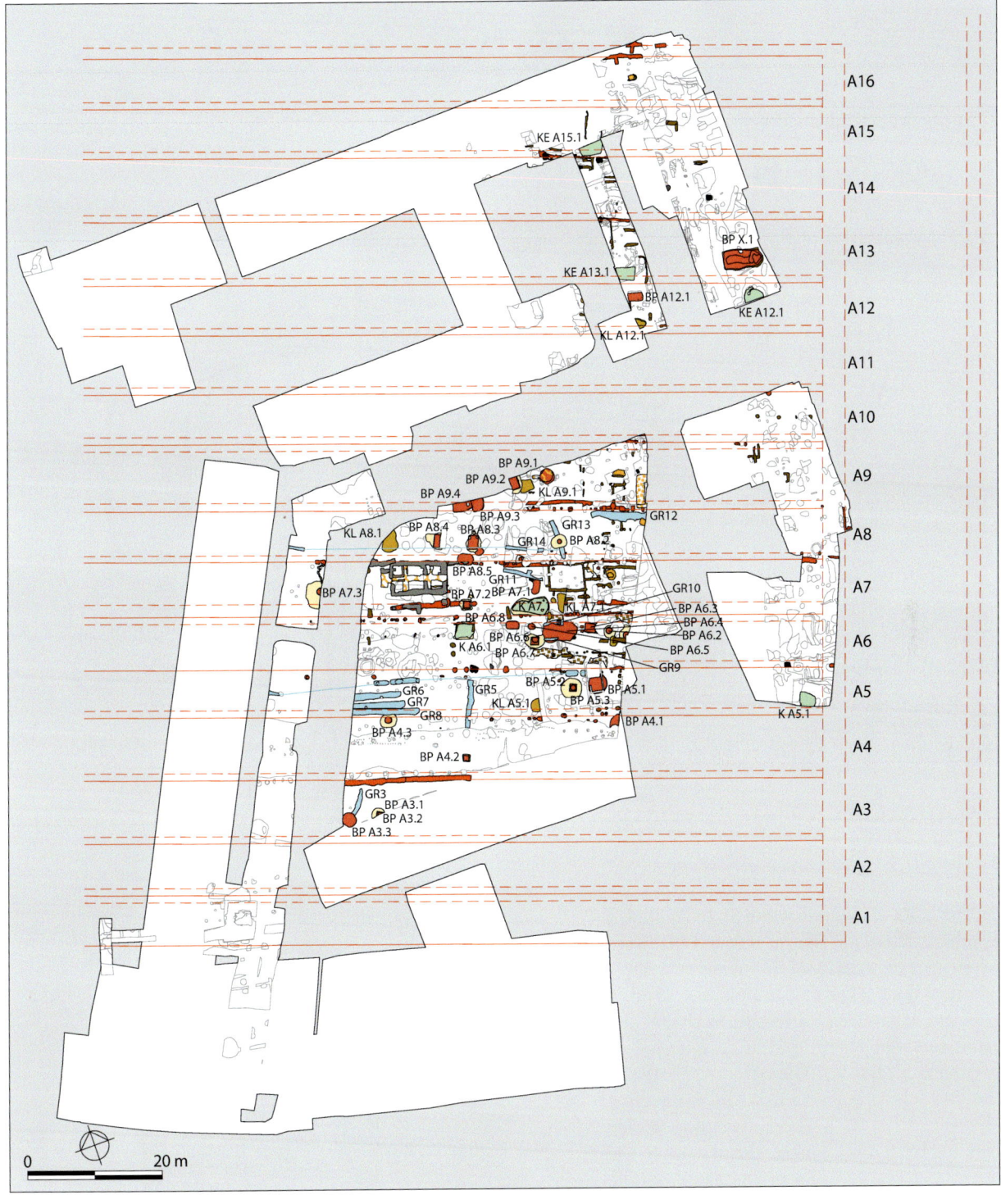

Figure 5. Feature map of the south-western corner of *Ulpia Noviomagus* (©TW/EH).

abandonment of the Roman town, this part of Nijmegen was no longer that densely inhabited, and although the ground level has remained virtually the same as that in Roman times, and 1st and 2nd century features in the area have been well-preserved, traces of toilet buildings should have been visible in some places at least. Since some privacy would have been appreciated, and thus it may be assumed that all plots will have had one or more privies, it seems that these small structures were constructed on lighter foundations of a different kind. Possibly, shallow ditches were dug, no deeper than the topsoil, which would make them archaeologically invisible. Another possibility is that these constructions were so light that they could be placed directly on the ground surface, without any foundation at all.

Figure 6. Reconstruction drawing of the toilet building on plot A4 on the St. Josephhof (© BLAN/RR).

3.2 Cesspits, cesspools and overflow ducts

On the plots excavated at the St. Josephhof, a number of cesspits have been found that were connected to a cesspool by means of a short overflow duct (Figures 7-8, Heirbaut 2010). A cesspool is similar to a cesspit in shape, and often cesspools were used as cesspits before becoming a cesspool. The difference between the two interpretations is in the connecting element, that is, a short declining gutter running from one element, the cesspit, to the other, the cesspool. From the same period, similar combinations of cesspit and cesspool are known from the military base at the Kops Plateau (Heirbaut in prep).

A remarkable aspect of the south-western part of *Ulpia Noviomagus* is that, with the exception of cesspit A8.3, these combined structures are completely lacking. A number of cesspits were joined together by a gutter conducting water, but in these cases the cesspits are at a greater distance from each other, and part of a system that may run across more than one parcel.

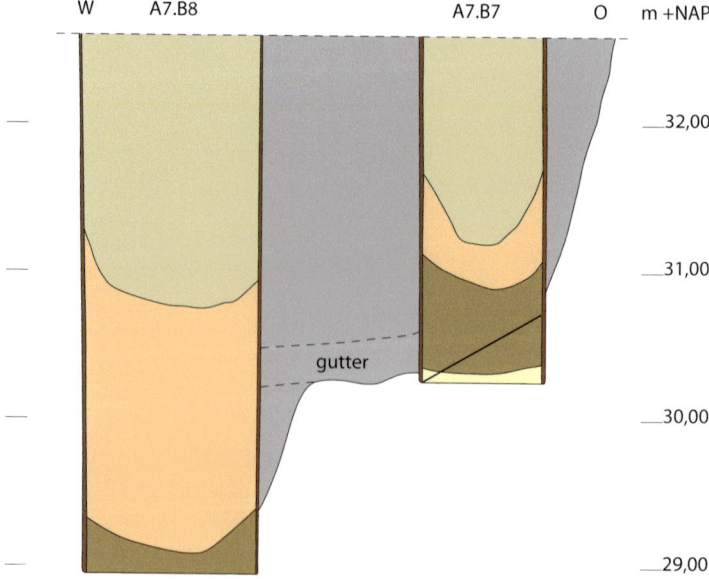

Figure 7. Reconstruction of cesspit A7.B7 (*Oppidum Batavorum*), which is connected to a cesspool via an overflow duct (©RM/EH).

Figure 8. Detail of the overflow duct on plot A7 (*Oppidum Batavorum*) (©BLAN).

However, the combination of cesspit and cesspool presupposes a short distance between the two.

As has been established above concerning the tendency to move cesspits from within the buildings to the outside from ca. AD 40, and the re-occurrence of in-house cesspits from the beginning of the 3rd century, the disappearance of cesspools may have been a temporal phenomenon, with the turning point at the end of the Flavian era.

3.3 Gutters for conducting water

These infrastructural elements have been observed during the period from the beginning of the 1st to the end of the 2nd century. On the 3rd century plots, no gutters conducting water have been found that can be connected to any cesspits. Gutters conducting water can be classified in two types, those serving a single parcel and those running across more plots.

The first group is by far the largest, both in *Oppidum Batavorum* as in *Ulpia Noviomagus*. The plots that have been excavated all show one or more gutters running over long distances and connecting several cesspits. In all instances, cesspits do not seem to have been in use simultaneously. On other plots, these gutters have not been observed, but that does not necessarily mean that they were not present. In some cases they may have been on the unexcavated parts of the plots. Many of these gutters for conducting water start inside the buildings. They were probably constructed below floor level, and most likely ran to the very rear of the plot. This made it easy, for example, to remove kitchen waste by flushing it down the drain. Only a few gutters conducting water that ran across more than one parcel have been found. Nowhere was such a gutter recorded to have started at the street front, but it is not impossible that some of them did. Both at the St. Josephhof and at the Rijn- en Lekstraat, the front parts of the houses have not been excavated, which makes it impossible to say anything definitive on this matter. The trenches along the modern Waterstraat have not been completely excavated. If these ditches were dug beneath floor level, they would only have become visible at a depth that was not reached during the excavations.

Both the sections and the reconstructed depth of these gutters relative to the reconstructed ground level during the Roman period show that the sewage system was completely underground. In contrast to the overflow ducts, the gutters for conducting water were not constructed from hollowed-out tree trunks. These are gutters with flat bottoms and straight sides, covered with wooden planks, which in turn were covered by soil. Similar examples are known from, e.g., Eschenz-Tasgetium (Jauch 1997: 17-19). Elsewhere, similar sewage networks have been found that had been constructed from roofing tiles (Henrotay 2007). The width of the ditches provides no indication for the width of the gutters; at some points, the ditches are over 1 m wide. The examples from Eschenz-Tasgetium were, on average, 30-50 cm in width. The wooden gutter from both inside and outside a barracks building in the *castellum* at Alphen aan den Rijn had a similar width (Bogaers and Haalebos 1987: 40-52). Most of the gutters at the St. Josephhof will have had comparable dimensions, but at one location indications were found for a gutter with a width of approximately 80 cm. The gutters identified at the Rijn – en Lekstraat (*Ulpia Noviomagus*) fit this picture.

The fact that the sewage system was largely underground means that the owners of the parcels suffered hardly any inconvenience from the (waste) water flowing through it. However, the underground position posed problems of maintenance; the system was less easily accessible in case of damage or clogging. Therefore, manholes will have been constructed in convenient places. During the St. Josephhof excavations, such manholes were found in some places adjacent to the ditches. However, such manholes have not been found anywhere in the south-western quarter of *Ulpia Noviomagus*. It remains unclear how these ditches were maintained. Where gutters were connected to cesspits or cesspools, the system could have been accessible from these structures. Still, it cannot be ruled out that, on occasion, simply digging a hole to solve a problem would have provided an easy solution.

The system described above does not require continuous flushing, but waste, whether from the kitchen or from cleaning the floors, would have had to be flushed out by a large amount of water. A smaller amount, such as that provided by emptying a couple of buckets, would not have been sufficient to conduct the waste from the house to the cesspit, ultimately resulting in clogging. Possibly, therefore, the system was connected to a system containing water that was running in the street.

4. Public Toilets

So far, no public lavatories have been located in *Oppidum Batavorum*. In *Ulpia Noviomagus*, too, only little is known. Although part of a bathing complex has been excavated, no traces of public lavatories were found there.

In the south-western *insula*, however, a workshop zone has been identified. This workshop dates to the early Flavian period – a time when there had not yet been any housing construction in this *insula*. It was oriented towards the *decumanus maximus*. Behind several pits filled with untempered clay, which belonged to the pottery established in this corner of the *insula*, was a very large cesspit (Figure 9). The size of the feature suggests a facility with seating capacity for a number of people. Lavatories like this were quite common in or near shops, taverns and industrial buildings. The

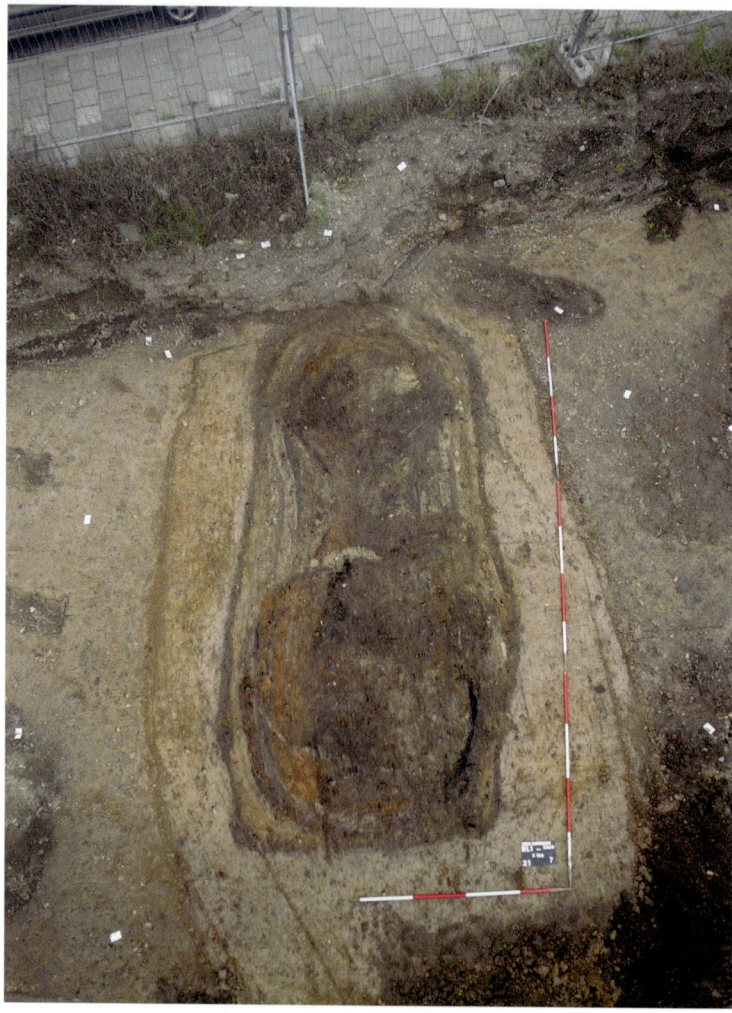

Figure 9a. View on and section of the public toilet in the south-western corner of *Ulpia Noviomagus* (©BLAN).

Figure 9b. View on and section of the public toilet in the south-western corner of *Ulpia Noviomagus* (©BLAN).

lavatories served the potters working here; whether it also catered for those working a little more to the east, in the workshops near the Maasplein just across the road, is unknown but quite possible.

Bibliography

Bogaers, J. E., Haalebos, J.K. 1987. Opgravingen te Alphen aan den Rijn in 1985 en 1986 *Westerheem* 36: 40-52.

Heirbaut, E. N. A. 2010. *Privé-toiletten uit Oppidum Batavorum. Opgravingen op de St. Josepphof in Nijmegen, 2* (Archeologische Berichten Nijmegen – Rapport 17). Nijmegen.

Heirbaut, E. N. A. (ed.) 2013. *De zuidwestelijke hoek van Ulpia Noviomagus in kaart gebracht. Deel 1. Resultaten van de opgravingscampagnes aan de Rijnstraat en Lekstraat in Nijmegen-West 2008-2010* (Archeologische Berichten Nijmegen – Rapport 41). Nijmegen.

Heirbaut, E. N. A. in prep: *De militaire bezetting van het Kops Plateau*. Nijmegen.

Henrotay, D. 2007. Le vicus d'Arlon. Renouvellement des connaissances, *Bulletin trimestriel de L'institut Archéologique du Luxembourg – Arlon* 1/2: 3-48.

Jansen, G. C. M. 1994. Romeinse privé-toiletten. In C.T. Waslander (ed.), *Latrines. Antieke toiletten – modern onderzoek:* 30-36. Meppel.

Jansen, G. C. M. 1997. Private toilets at Pompeii. Appearance and operation. In S. E. Bon and R. Jones, (eds), *Sequence and Space in* Pompeii: 121-134. Oxford.

Jauch, V. 1997. *Eschenz-Tasgetium. Römische Abwasserkanäle und Latrinen* (Archäologie im Thurgau 5). Frauenfeld.

van Enckevort, H., 2010. Sporen uit de vroeg-Romeinse tijd. In H. van Enckevort and E. N. A. Heirbaut, (eds), *Opkomst en ondergang van Oppidum Batavorum, hoofdplaats van de Bataven. Opgravingen op de St. Josephhof in Nijmegen 1* (Archeologische Berichten Nijmegen – Rapport 16): 241-252.

van Enckevort, H. and Heirbaut, E. N. A. (eds) 2010. *Opkomst en ondergang van Oppidum Batavorum, hoofdplaats van de Bataven. Opgravingen op de St. Josephhof in Nijmegen 1* (Archeologische Berichten Nijmegen – Rapport 16): 50-51.

Arlon, apport des découvertes récentes dans le vicus à l'examen des latrines gallo-romaines

Denis Henrotay

1. Arlon dans la Trévirie

L'agglomération antique d'Arlon est structurée autour du croisement de deux chaussées importantes : la Reims – Trèves et la Metz – Tongres dans la partie occidentale de la cité des Trévires en Belgique seconde. Sa topographie est marquée par un signal fort : une colline qui sera fortifiée lors de l'Antiquité tardive. Le cimetière du *Hochgericht* a révélé de nombreuses sépultures dont les plus anciennes remontent à le première décennie de notre ère. Au IIIe siècle, l'itinéraire d'Antonin mentionne l'*Orolauno uicus*.

Des recherches archéologiques récentes ont permis la mise au jour de plusieurs quartiers artisanaux dans le *uicus* d'Arlon, qui présente au Haut Empire la physionomie d'un village rue. Les maisons de plans allongés sont implantées perpendiculairement aux voiries. Un espace ou venelle sépare chaque habitation et récolte les eaux de pluie qui sont ensuite dirigées vers la zone arrière dans un égout. Les bâtiments de ce quartier sont construits dès l'époque flavienne suivant la technique du pan-de-bois reposant sur un soubassement de pierres. L'accès est souvent ménagé dans la partie centrale de l'édifice. Deux pièces en façade encadrent un couloir central qui mène à la partie arrière. Les dimensions sont approximatives de 11m de largeur pour 22 à 24 m de longueur.

Plusieurs latrines ont déjà été mises au jour dans la région arlonaise. Il s'agit de structures fonctionnant avec des égouts. Dès 1907 (Loes 1909), des latrines simples à égout sur trois côtés perpendiculaires ont été découvertes dans les thermes du *uicus*. Une rigole creusée dans de longues pierres calcaire permet de restituer le fonctionnement des lieux. Les dimensions sont de 4,5m par 5m. L'alimentation en eau des latrines provenait de la piscine voisine.

La villa rurale de Mageroy à Habay-la-Vieille est implantée à quelques kilomètres à l'ouest d'Arlon (Zeippen 2009). On y a trouvé des latrines simples à égout sur le long côté qui étaient alimentées par une canalisation en bois. De dimensions plus modestes (16 m^2), elles sont bâties à l'angle opposé à la partie thermale durant la phase tardive de l'occupation de la villa.

2. L'ancien site industriel Neu, rue Goffaux

Les anciens établissements Neu, édifiés en 1948, étaient jusqu'il a peu un site industriel désaffecté en bordure des voies de chemin de fer. Plusieurs tranchées de sondage y ont révélé le potentiel archéologique de l'ensemble de la zone couvrant plus de 17.000 m^2. Plusieurs facteurs ont favorisé l'accumulation de tourbe sur le site : un fond de vallée très peu pentu (1%), la proximité immédiate d'une petite rivière, la Semois, et la nature argileuse du sous-sol. Les épisodes d'extension du lit du cours d'eau ont été mis en évidence avant, pendant et après l'occupation antique. Ce milieu humide s'est révélé excellent pour la conservation des éléments d'origine organique.

En 2003, un projet de construction d'un bassin d'orage permettant de réguler la Semois lors des forts afflux d'eau liés aux pluies et de résoudre les problèmes récurrents d'inondation, nous a fourni l'opportunité d'explorer 1.200 m^2 de ce site. Le décapage, centré sur la zone menacée dans l'immédiat par la construction du bassin, a rapidement révélé la présence de cinq parcelles bâties. La partie excavée (65 x 19 m) formait une coupe perpendiculaire à la vallée suivant un axe nord-sud. La partie sud, la plus profonde et la plus humide était également la mieux conservée. Les bâtiments dans leur ensemble ont été détruits à la fin du IIIe siècle. Cinq puits et latrines furent comblés à cette époque et formaient les plus beaux ensembles clos. Des objets en bois et en cuir y étaient parfaitement conservés. L'occupation du site s'est poursuivie durant le IVe siècle. Les couches de démolition de la fin du IIIe siècle et les bétons de sol ont été percés par de nombreux poteaux. Les murs arasés du Haut-Empire ont parfois été réutilisés et surmontés par des maçonneries en grès ferrugineux posées sans liant, employant des matériaux totalement absents auparavant. Le parcellaire semble donc avoir perduré. Toutefois, ce sont ces niveaux tardifs qui ont subi l'érosion la plus forte. Les traces d'activité disparaissent après le IVe siècle. Par la suite, le ruisseau présentera une nouvelle période d'extension et recouvre le site d'une couche de tourbe.

En 2004 et en 2006 (Henrotay 2006), les parties arrières de deux autres bâtiments furent révélées. Elles étaient enfouies à faible profondeur dans un terrain en place

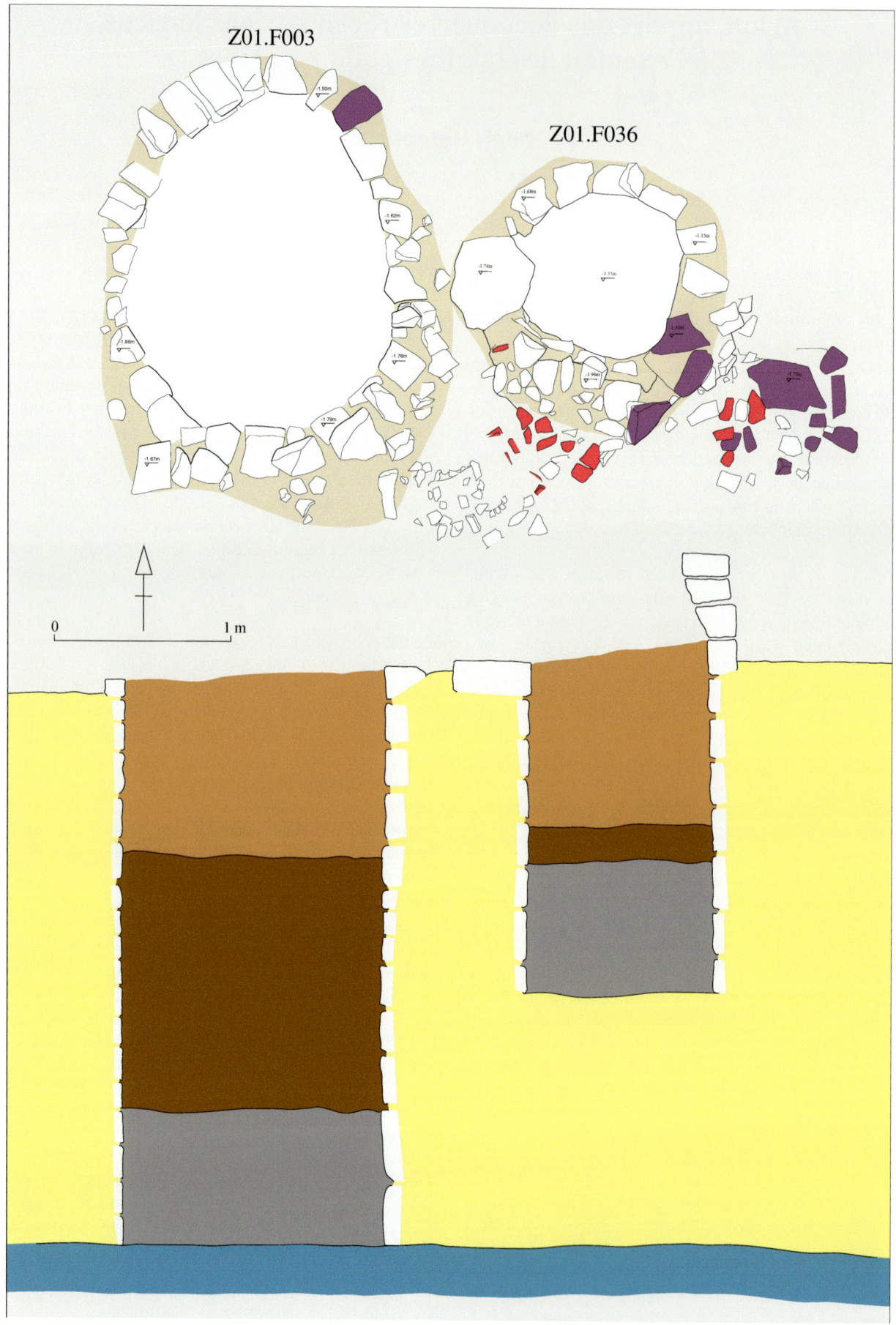

Figure 1.

très argileux. Seules furent mises au jour les parties excavées. Les celliers ou caves ont ainsi constamment été maintenus en milieu humide, ce qui a favorisé la conservation exceptionnelle des planchers et des aménagements en bois. Leur positionnement au nord de la partie fouillée en 2003 et leur implantation selon un axe nord-sud, c'est-à-dire perpendiculaire à l'axe est-ouest des cinq premières parcelles démontrent qu'à partir de cet endroit le parcellaire antique forme un angle d'îlot. Les deux bâtiments s'alignent sur une voirie perpendiculaire à l'axe de la voie Metz-Tongres ou de l'actuelle rue Goffaux.

L'étude de l'abondant matériel céramique (Hannut/Henrotay 2006) a révélé la présence de 3.619 vases pour un total de 31.732 tessons. Elle a permis d'appréhender l'approvisionnement céramique du *uicus* pour la période allant de la fin du Ier au IVe siècle. Aucun pot de chambre ou *lasanum* n'a été découvert dans l'ensemble des sites arlonais comme ce fut le cas dans le *uicus* de Liberchies (Brulet *et al.* 2001).

3. Les données archéologiques concernant la mise au jour de latrines

Deux ensembles de latrines ont été découverts sur le site Neu. Un premier groupe consiste en un alignement de quatre puits qui matérialise une limite parcellaire. Groupés par deux, ils présentaient des caractéristiques et des fonctions différentes à l'origine. De plan circulaire, les puits à eau sont beaucoup plus larges et plus profonds (3,40m) que les latrines (Figure1). Ils récoltent les eaux qui ruissèlent sur le banc de roche en place. Le débit est abondant. La profondeur des latrines quand à elles n'excède pas 1,80 m et n'atteint pas le niveau de la nappe phréatique. Cependant la proximité immédiate des latrines a dû entraîner la contamination des puits à eau et motiver leur reconversion en latrines comme l'atteste l'étude des remplissages d'utilisation. Le matériel archéologique mis au jour dans le comblement de démolition des quatre puits et latrines démontre leur utilisation et leur abandon simultané après une destruction générale du *uicus* dans les années 270-280. Le contexte anaérobie du site à proximité de la rivière a favorisé la conservation des matières organiques. De nombreux restes végétaux ont été découverts, parmi ceux-ci les taxons d'une céréale (orge vêtue), de 14 espèces de fruits, de 10 plantes condimentaires ou légumes et de 34 plantes sauvages (Derreumaux 2009). Des fragments de 4 taxons de bryophytes ont également été mis au jour. Cependant, le matériel n'est composé que de fragments de gamétophyte, il manque les sporophytes dont la forme entre rapidement en jeu dans les clefs de détermination. La présence de plusieurs espèces différentes de mousses suggère une collecte de ces végétaux utilisés pour l'hygiène corporelle. Les habitants avaient probablement recours à cette ressource locale abondante pour suppléer à l'absence d'éponge marine plus coûteuse. L'analyse micro-archéologique (Laurent 2008) a révélé dans tous les puits la présence de coprolithes mais aussi de pupes de larve de mouches dont certaines étaient minéralisées. Ce faisceau d'indices nous permet d'attribuer la fonction de latrines aux quatre puits cuvelés de l'alignement parcellaire.

En 2006, un nouveau projet immobilier mettait en péril une zone contiguë à celle excavée pour le bassin d'orage. Nous y avons découvert un cloaque construit dans le sens de la pente selon un axe nord-sud. Il délimite l'arrière des parcelles des bâtiments d'habitation découverts en 2003. L'espace du jardin est ouvert sur une longueur de plus de 24 m. Les analyses polliniques révèlent une zone occupée par une strate herbacée correspondant à une friche et prairie pâturée (Defgnée 2009). Les parois de cet égout étaient aménagées au moyen de piquets de chêne. Un plancher, bien conservé par endroits, formait un couvercle. Ce cloaque a été daté grâce à la dendrochronologie des années 154+/-10. Des canalisations en bois et en pierres amenaient les eaux de pluie des toitures à cette évacuation vers la rivière. Elles ne sont pas toutes contemporaines et reflètent donc des réfections. Elles prolongent les alignements parcellaires matérialisés par les espaces récoltant les eaux pluviales entre les maisons. La période de fonctionnement de la structure se termine à la fin du IIe siècle ou au tout début du IIIe siècle, comme l'atteste le matériel céramique accumulé dans son comblement. La faible pente du fond de vallée est propice à l'accumulation des sédiments. C'est la cause de l'engorgement et l'abandon de l'égout. Les niveaux de circulation dans les habitations ont été exhaussés de près de 70 cm en deux siècles. Un amas de tuiles recouvrait une section du cloaque. A cet endroit plusieurs piquets en chêne renforçaient la canalisation et dépassaient du conduit. Il s'agit probablement de latrines sur égout. Il est intéressant de signaler la présence toute proche d'un petit bac en pierre de récupération (probablement d'origine funéraire) à la fin d'une évacuation de eaux de pluie (Figure 5). Son rôle était probablement à mettre en rapport avec les ablutions hygiéniques. Un second bassin en pierre a été découvert à l'emplacement d'une autre limite parcellaire mais, à cet endroit, le cloaque était détruit par des structures contemporaines.

Après l'abandon de l'égout, durant tout le IIIe siècle, les latrines sont uniquement constituées de petits puits en pierres régulièrement reconstruits. Ces structures successives respectent l'organisation parcellaire. Chaque maison possédait ses latrines individuelles. Le remplissage des structures était homogène, très organique et contenait peu d'artefacts. Les latrines consistaient en des puits aménagés au moyen de pierres sèches et dont la profondeur varie de 0,90 à 1,20m. Un plancher couvrait partiellement les fosses circulaires, l'espace restant marquait l'emplacement

Figure 2. Bac en pierre disposé à la fin d'une canalisation d'eau de pluie à proximité des latrines construites sur l'égoüt en bois

Figure 3. Le plancher en chêne des latrines en forme de puits est conservé. Les canalisations en pierre ammènent les eaux de pluie à l'arrière de la parcelle.

Figure 4. Latrines de plan rectangulaire avec les poteaux de la superstructure.

d'une banquette en bois disparue (Figure 3). Ce type de puits peu profond a été rencontré à de très nombreuses reprises à l'arrière des habitations dans l'agglomération gallo-romaine de Mageroux à Saint-Mard située à une trentaine de kilomètres d'Arlon sur le territoire actuel de la ville de Virton. Ces fosses délimitées par des murets de pierres posées sans mortier y ont été interprétées comme de citernes (Cahen *et al.* 1994). Le contexte de découverte y est moins favorable que dans la vallée de la Semois ; aucun élément organique n'y a été mis au jour.

A une seule reprise, une fosse de plan rectangulaire a été mise en évidence. Les parois de la fosse creusée dans l'argile sont simplement renforcées par six solides pieux qui soutiennent le plancher sur lequel circulaient les utilisateurs. Un espace libre de plancher marque à nouveau l'emplacement de la banquette. Une superstructure en bois, dont on a retrouvé les six poteaux porteurs (Figure 4), englobait le tout et permettait aux utilisateurs de pratiquer les lieux à l'abri des intempéries et en toute discrétion.

Outre ces données architecturales et spatiales, les remplissages d'utilisation des structures interprétées lors de la fouille comme étant des latrines mais aussi d'autres fosses à vocation artisanales ont fait l'objet d'études pour y mettre en évidence les restes éventuels

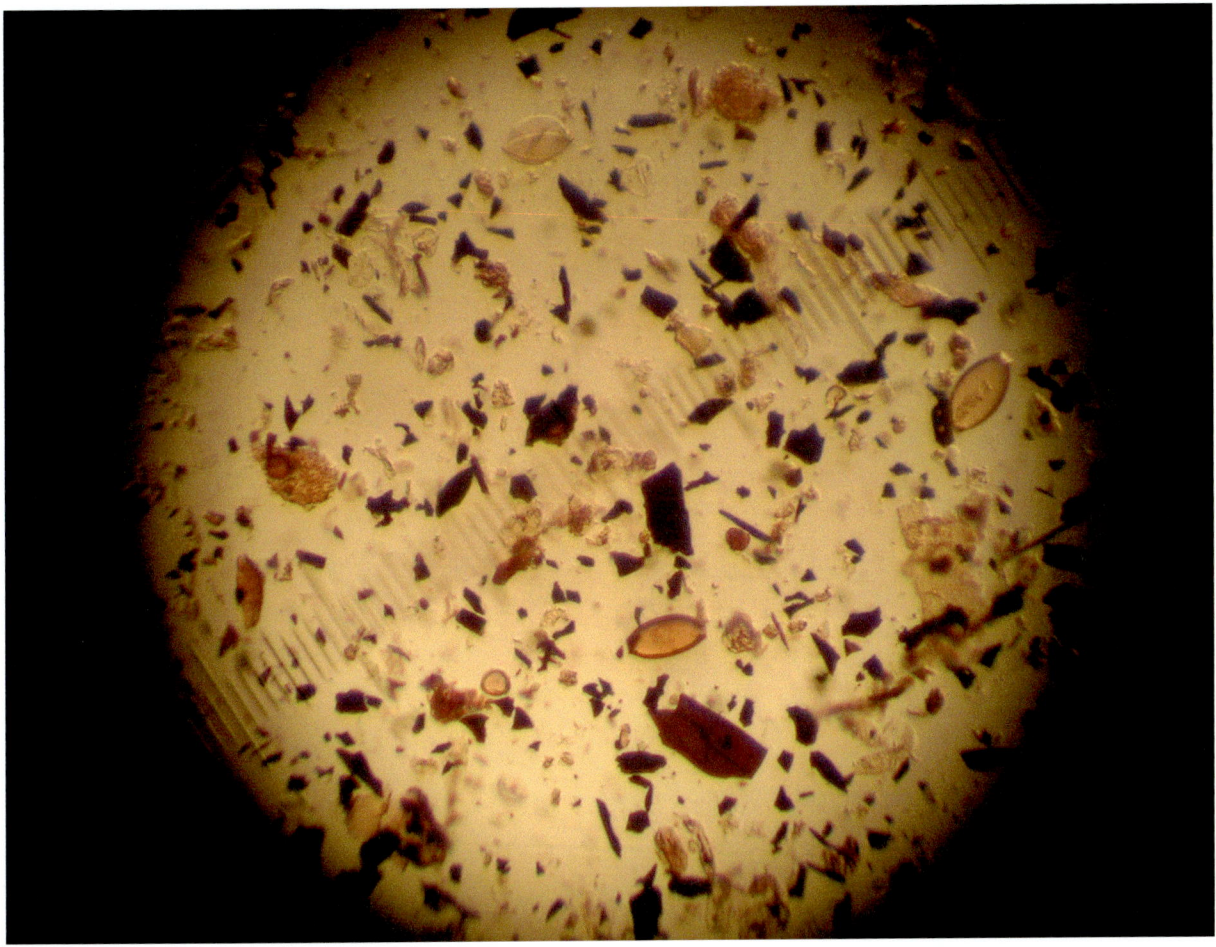

Figure 5. Palynofaciès des latrines : pollens de céréales et oeufs de Tricuris

de pollens et de parasites. C'est ainsi que dans les latrines, de nombreux pollens de céréales (Defgnée 2009) représentant plus de 15,62% de l'échantillon étaient associés aux très nombreux œufs d'helminthes (*Tricuris* et *Ascaris*) (Figure 2). Cette présence massive de céréales trahit la consommation de la bouillie ou *puls*, substitut à la consommation du pain. Cette préparation est l'aliment de base de la majeure partie de la population gallo-romaine. Des pollens de vigne (*Vitis vinifera*), de carottes (*Daucus carota*), de chénopode *blanc* (*Chenopodium album*), d'anis vert (Pimpinelle anisum), de sureau (Sambucus), de mûres sauvage (Rubus) et de sauge (Salvia) ont également été identifiés. D'autres fosses découvertes sur ce chantier ont fait l'objet d'analyses polliniques mais la présence très faible de pollens de céréales (1,45%) confirme les observations réalisées sur le terrain. Il ne s'agit pas de latrines.

4. Conclusion

Dans ce quartier artisanal, un grand soin a été apporté à l'évacuation des eaux pluviales. Ces eaux alimentaient un égout et emportaient les matériaux indésirables. Plus tard au IIIe siècle, la concentration des habitations a entraîné la reconversion des puits à eau en latrines. Chaque habitation possédait sa propre installation sur puits qu'elle renouvelait régulièrement. Ce sont les analyses des pollens, des parasites et la micro-archéologie qui ont été déterminantes pour prouver avec certitude la fonction exactes des structures.

Bibliographie

Bouet, A. 2009. *Les latrines dans les provinces gauloises, germaniques et alpines* (Gallia, supplément 59). Paris.

Brulet, R., Dewert, J.-P. and Vilvorder, F. (eds), 2001. *Liberchies IV, Vicus gallo-romain, Travail de rivière, fouilles du Musée de Nivelles 1986/87 et 1991/97*: 335-337. Louvain-la-Neuve.

Cahen A., Clausse R., Gautier A., Lallemand J., Lambert-Henricot C. et Massart CL., 1994. *Un quartier artisanal de l'agglomération gallo-romaine de Saint-Mard (Virton)* (Etudes et documents 1). Namur.

Derreumaux M. 2009. La carpologie ou les macrorestes végétaux du uicus d'Arlon. Catalogue de l'exposition Les experts à Arlon, Autopsie d'un uicus. *Bulletin de l'Institut archéologique du Luxembourg* 86, 1-2: 147-150.

Defgnée A. 2009. La palynologie au service du uicus d'Arlon. Catalogue de l'exposition Les experts à

Arlon, Autopsie d'un uicus. *Bulletin de l'Institut archéologique du Luxembourg* 86, 1-2: 151-153.

Hannut, F. et Henrotay, D. 2006. Le mobilier céramique des IIe et IIIe siècles du site 'Neu' à Arlon/Orolaunum (province de Luxembourg, Belgique). *Eléments pour la définition du faciès céramique de la partie occidentale du territoire trévire*. Actes du Congrès de Pézenas, Société Française d'Etude de la Céramique Antique en Gaule: 287-339. Saint-Paul-Trois-Châteaux.

Henrotay, D. 2004. Arlon/Arlon : vestiges du vicus. *Chronique de l'Archéologie wallonne* 12: 177-178.

Henrotay, D. 2006. Arlon/Arlon : habitat gallo-romain sous l'ancien site industriel Neu. *Chronique de l'Archéologie wallonne* 13: 201-202.

Henrotay, D. 2007. Le vicus d'Arlon : renouvellement des connaissances. *Bulletin de l'Institut archéologique du Luxembourg* 1/2: 3-48.

Henrotay D. 2008. Arlon/Arlon : rue Goffaux, vestiges gallo-romains découverts sous l'ancien site industriel Neu, campagne 2006. *Chronique de l'Archéologie wallonne* 15: 168-169.

Laurent Ch. 2008. *Arlon, site Neu : aperçu micro-archéologique et carpologique, rapport de convention*. Archives Service Public de Wallonie.

Loes, F. 1909. Découvertes romaines faites à Arlon en 1907. *Annales du XXe Congrès*: 253-268. Liège.

Zeippen, L. 2009. Habay, Habay-la-Vieille, La villa de Mageroy. In R. Brulet (ed.), *Les Romains en Wallonie*: 469-474. Bruxelles.

A Roman latrine near St. Kolumba in Cologne and its remarkable contents

Michael Dodt

Introduction

Pits found during excavations in the Roman layers of Colonia Claudia Ara Agrippinensium often can be identified as latrines. They are more or less round, with a straight section like a shaft, which distinguishes them from clay pits. An important criterion distinguishing them from wells is their depth, as Roman wells in Cologne reach down into water-carrying layers at c. 38 m above Normalnull[1] (see Hellenkemper 1986: 206–208). The wells were build in spite of the inferior quality of the water in them even during the operation of the aqueducts bringing water from the Vorgebirge and the Eifel southwest of Cologne during the High Empire.

Rubbish pits can have the same form as latrines as well, especially as both latrines and clay pits were re-used as rubbish pits, like the latrine presented here. Particular characteristics of latrines are traces of faeces in the lower layers of the filling, which stand out as greenish streaks. They contain remains of either humus, or – with better preservation – of plants such as fruit pits, nuts etc. In the upper layers, these pits are filled with household refuse, pottery and such.

Contrary to the medieval latrines in the area, which were emptied from time to time, the Roman latrines were relatively short-lived.[2] The moment of filling as dated by the finds thus is close to the first digging of the pit.

This construction differs from Roman latrines above brick channels, which mainly occur in bath buildings, where they are flushed with the accumulated 'grey' wastewater. This is especially visible in the private bath of the *Villa rustica* of Cologne-Braunsfeld, which is unfortunately less well preserved otherwise (Bouet 2009: 372; Dodt 2001: 315–319). It also differs from the latrine connected to the channel system behind the house of the Dionysos-mosaic (Insula I/J 1, see Precht 1971: Beilage 1), and from a latrine situated just 30 m to the southeast from the site of the latrine under the northern graveyard of St. Kolumba presented here (Figure 1, Excavation RGM Köln FB 2001.021 channel 124/133 resp. Fundbericht 74.03 at the western end of channel 91, on the other side of wall 71).

Latrine with channel system

Before we get to the shaft-like latrine, I would like to briefly describe this 'basin' connected to the channel-system. Like the channels, it is constructed from bricks with a firm mortar and a bottom made from whole *tegulae* or *lateres*. Channel 91, which is covered with large slabs of greywacke (a variation of sandstone), comes through the NS-wall no. 71 from the east and turns at a right angle northwards about 0.90 m to the west of the wall, where it forms a 0.70 m wide basin. The cover of channel 91 in wall 71 is made of a former threshold, with the holes for the hinge and bar still visible. A medieval cellar wall has disturbed the channel's further course to the north (Excavation RGM Köln FB 74.03, 'basin' at channel 91 and FB 2001.021 context 129). Because of the level of the bottom of the channel at 49.18 above Normalnull and the connection with wall no. 71, the associated occupation layer is at 49.93 m above Normalnull. This layer can be assigned to period IV, dating into the second half of the 3rd and the 4th century AD. Dating the construction of the latrine or channel 91 is impossible, as the finds from the filling do not date to the abandonment of the construction. The channel system connected to this latrine was build from stones and bricks and forms a typical feature of the later Roman building period of one of the houses near St. Kolumba. An increase in channels during Late Antiquity has been observed at other places in Roman Cologne, e.g. in the southern *suburbium* (see Dodt 2005).

Cesspit-latrine

In contrast to this installation, the shaft-like latrine 95 of excavation FB 2003.017 was constructed from wood, which is fairly easy to recognize. In addition to that, the filling contained exceptional and well-dated finds, which – together with the construction – characterize the pit. This was only recognized during the analysis of

---

[1] A vertical measurement referring to the height above mean sea level (see entry 'Normalnull' in Wikipedia).
[2] Medieval latrines from the excavations at St. Kolumba are circular, built from stone and date from the 13th to the 15th century AD (RGM Köln FB 2001.021 contexts 21, 47 and 161 and FB 2003.017 context 11). The find spots are immediately to the north of the outer graveyard of the church, which reached the apex of its building history with the late Gothic extension of between 1460 and 1530 (Bau V, see Dodt and Seiler, forthcoming).

Figure 1.

the documentation of the St. Kolumba excavation. This contribution thus originates in an accidental find and not a systematic comparison of all remaining Roman latrines of the CCAA. During the excavations, this latrine had been interpreted as a pit and afterwards time constraints prevented a re-evaluation and adjustment in the documentation. The analysis took place at the same time as the analysis of the excavation in the *vicus* of Bonn, where a better-preserved example of a latrine of the same construction type was found (see Andrikopoulou-Strack *et al.* in this volume).

The latrine with the running number 95 was excavated under a Late Roman cellar during the excavations preceding the erection of the new museum at St. Kolumba in 2003. The Roman contexts of the excavation had been much disturbed by modern building work, especially directly after WWII. Only a small area was left undisturbed, it contained the eastern wall of a later Roman cellar and a part of the western wall connected to it (excavation RGM Köln FB 2003.017, object-no. 24/102 und 25/96). Any architectural connection to the earlier houses of the 1st and 2nd century AD is difficult to reconstruct, not least because of the later Roman interventions. The site is in a mixed *insula* of houses and workshops between the 1st and 2nd *cardo* west of the *Cardo Maximus* and the 2nd and 3rd *decumanus* from the north, thus *Insula* F3/4 after the common classification (see Hellenkemper 1975: 785).

The Roman settlement in this quarter starts in the first half of the 1st century AD and runs through to the 5th century AD. A continuation beyond the 5th century is represented by the addition of an apse onto a Roman room in the 7th century, which became the nucleus of the oldest parish church of Cologne (see Seiler 1977: 97–119; Seiler 1989: 146–157). The plots of the Roman *insula* have an EW orientation in this area, whose deep enclosing walls on the north and south sides could be excavated (Excavation RGM Köln FB 2001.021; see Dodt and Seiler forthcoming). The eastern street-facing fronts of the houses and their rear ends to the west could not be found, although the course of the *cardo* immediately to the northwest of St. Kolumba is known. The plots are rectangular and long, the houses are so-called strip-houses (see Dodt 2002: 640–642; Dodt 2005: 722–732; Precht 2002: 181–198). As a rule, such plots were only built up in the front part near the street, with the middle and rear part often containing open courtyards or gardens, around which rooms are grouped, such as a bath in a house of *insula* 19 in Colonia Ulpia Traiana (Xanten, see Dodt2003: 324–327, fig. 131). The latrine is set into the plot c. 43 m west of the 1st *cardo,* 52,5 m east of the 2nd *cardo* and 30 m south of the 2nd *decumanus* south of the northern wall. The *cardines* of the CCAA run almost precisely in north-southern direction and the 2nd *cardo* lies almost under today's Nord-Süd-Fahrt/Tunisstraße. The 2nd *decumanus* running in EW direction almost equals today's Breite

Figure 2.

Figure 3.

Straße/Minoritenstraße. Because of this position, it is likely that the latrine was situated in the rear part of the plot. The middle part of the plot has been excavated in a more coherent excavation in 2001, in which the boundary walls to a northern and a southern plot were discovered (Excavation RGM Köln 2001.021, contexts 56/96/97/102 und 197/208). The building technique of these three plots differ wildly: in the southern plot, basins and channel systems were build, while the middle plot had been build up with simple living spaces in a mixed technique of *opus caementicium* and timber-frame with wooden floors. Timber-framed walls and floors with floorboards of fir have been preserved by a big conflagration. One of the walls was still standing upright and thus was the tallest standing Roman wall in Cologne at that point (see Dodt and Seiler forthcoming). Some of the rooms in the northern plot have hypocaust heating. Both a hypocaust with pillars (room 23) and a channel-hypocaust (room 46) have been found. While the latter is usually seen as the later technique, both were in use at the same time in this case.

Pit 95 of excavation FB 2003.017 lies west of the northernmost of these three Roman plots, under the northern wall (24) of a large Roman cellar (Figure 2). The bottom is at 46.48 m above Normalnull and it goes down 0.48 m into the natural sand under the cellar wall (Figure 3). The natural sand in the area of the cellar lies deeper than to the east of the cellar. On top of the sand lies natural clay with an upper edge at c. 48.80 m above Normalnull, which had been removed at the site of the cellar.

The shaft of the latrine had originally been dug from the upper edge of the natural clay at 48.80 m and the latrine thus had a depth of almost 2.40 m. The pit is 1.30 m wide, with a flat bottom and a steep western wall (the eastern wall could not be documented). The lower layer of filling consists of dark brown sand. Above that lies a layer of olive-brownish, sandy clay with small pieces of mortar, and above that, a layer of mortar. The finds label shows, that the top layer is formed by a stone cover. The dark brown colouring of the pit's walls is a result of the former wet contents of the latrine often occurring in latrines. Other indicators are the colouring of the lower layers of the filling as

Figure 4a/b.

well as the crusts on some fragments of pottery (Figure 4a/b). Due to time constrains, the pit could not be fully excavated and documented and no sample of the lower filling was taken. While latrine 95 has the diameter of a well, with a depth of 46.48 m above Normalnull it does not reach the depth of the groundwater table (see Hellenkemper 1992: 208). At the excavation of a Roman well found nearby under the Romanesque cellar in the middle part of the excavation to the east of this one, the bottom was not reached at even c. 46.00 m above Normalnull (Excavation RGM Köln FB 2001.021, context 219). A ring of tuff ashlars in circular segments attests to its use as a well. The rounded and drawn out corners of the southern part of latrine 95 (Figure 3) form an indication of its construction and function (Excavation RGM Köln FB 2003.017). These corners obviously were made by round posts, such as the ones found in a better-preserved latrine in the *vicus* of Bonn (see Andrikopoulou-Strack *et al.* in this volume). In both cases, they indicate a construction of a seat made of a board or beam, mounted between posts. After the latrine had been given up, the posts were drawn out and their holes filled up with the contents of the pit, so that the greenish-brown edges from the pit and the outer sides of the postholes fused. Finally, the filled up pit was covered with the northern cellar wall 24, which was also set on the older wall 102, just like the southern cellar wall 25 had been built on the NS-wall 96. Both of the older walls are badly preserved and both are either connected to latrine 95 or take it into consideration, so that they must have existed at the same time. The latrine must have been abandoned with the building of the cellar at he latest. While the filling dates the abandonment of the latrine round 100 AD (a slow filling over the course of several decades can probably be excluded), a *terminus ante quem* for the abandonment of the walls 96 and 102 is given by the building of the cellar in the second half of the 2nd century AD, as the find of a rim of a TS-Kragenschüssel type Dragendorff 38 demonstrates. It was discovered with remains of mortar clinging to it in the corner of walls 24/25 under wall 25. This indicates that the abandonment of the latrine and the walls may not have been causally connected. Because of the interpretation of pit 95 as latrine, an interpretation of the walls 96 and 102 as cellar walls can be discounted.

The finds from the cesspit-latrine

The pottery from the filling of the latrine – mainly a decorated Terra Sigillata bowl type Dragendorff 37, a Terra Sigillata plate Dragendorff 15/17, a Terra Sigillata bowl Dragendorff 27, a Terra Nigra bowl Deru B16 and a cup Hofheim 82 (Figure 4a/b, plus a fragment of a smooth white vessel on a 'stone cover') – date the abandonment and subsequent filling of the pit round 100 AD. But there is more from the pit: some pigment balls of common madder (Figure 5), several fragments of glass vessels and a brick fragment with glazing, which probably came from an oven (Figure 4b). Some finds indicate at least one workshop in the immediate surroundings of the latrine: the glazed fragment of brick is likely to have come from the oven of a glass manufacture, which may also have produced the vessels of clear and blueish glass. Because of the high amount of pigment found, a use as make-up was at first excluded. Why the pigments were preserved at this spot could not be determined at the excavation. The use in wall painting was also excluded because of the colour tone of the pigment found. Red pigments for wall painting in Roman Cologne are made from ochre (see Noll *et al.* 1972/73: 80-81. Information kindly provided by Prof. Renate Thomas, RGM). A micro X-ray fluorescence analysis by the Rathgen-Forschungslabor in Berlin showed the pigments to be a madder paint (analysis rapport no. 213_092906 by A. Unger and S. Schwerdtfeger). In antiquity, common madder (*rubia tinctorum*) was mainly used in the dying of cloth (see Schweppe 1993: 232-234; for its use in wall painting see Eibner 1970: 181). The laboratory could also confirm the lack of substrate, which would be logical with the high amount of calcium. A substrate of gypsum or chalk would chemically contradict the use for the dying of cloth as well. If we consider the technical process of dying, there is a discrepancy in the use of the pigments. As a rule, the Romans cooked the unprocessed wool with the leafs and roots of the madder plant, which are the natural transmitters of the pigment. The concrete function of the pigment of the common madder is yet to be determined (see Puybaret *et al.* 2008: 185–193. I thank Dr. Annette Paetz gen. Schiek for the information). On the other hand, Pseudopurpurin and Purpurin, the pigments found in the latrine, are both elements of madder paint also discovered during the analysis of pinkish-red pigments found in glass *balsamaria* for cosmetic purposes in Cologne, Trier and other Roman places (see exhibition Le bain et le miroir 2009: 132–134 Cl 55 u. 57). The analysis of pink remains from glass balls found in Nijmegen and Vechten (the Netherlands) and in Moyland (Kreis Kleve, Germany) determined, that the colouring agent pupurine (or common madder) was always found in connection with gypsum (see Hottentot and van Lith 2006: 187–189 Nr. 5, 8 u. 9). As no binding agent in the form of starch or egg white, glue, resin or rubber could be found and also no oils or other fats, the contents must have been used as a loose powder for sprinkling. The samples from St. Kolumba and the samples described by Hottentot and van Lith were analysed with the same method (high-performance liquid chromatography). The latter samples contained common madder, a common red pigment in antiquity. This was the first time common madder was found in closed Roman vessels. The vessels from Nijmegen, Velsen and Moyland also contained large amounts of henna. Hottentot and van Lith (2006: 189) also mention red samples from Athens and Corinth

Figure 5.

on a basis of aluminium oxide. If we consider the size of these *balsamaria* and the amount of pigment they must have contained, the amount of pigment found in latrine 95 of St. Kolumba may conceivably also be interpreted as make-up. However, common madder has been found only rarely in Roman wall painting (I thank Prof. Dr. Renate Thomas and Dr. Nicole Riedl for this information). This rather sizeable amount of pigment in the balls will surely not have been thrown into the latrine on purpose.

The analysis of soil samples from pit 95 did not include specific indicators for a Roman latrine (e. g. caraway seeds), but this cannot be interpreted as a proof for the opposite as the sample was taken from the upper layers (where the pigments were also found). This probably was a layer of refuse filled into the latrine after the abandonment (analysis carried out by Silke Schamuhn, University of Cologne in the framework of a larger project on archaeobotanics).

The remarkable ground plan of the pit reveals the construction of the latrine above ground. It also raises the question of the construction of those latrines in Cologne that are without discernible remains of postholes or walls. The argument that these were simply not noted during excavation can be refuted,

as indicators of this construction were not noted in both old excavations and more recent excavations. The circumstances of the discovery of this latrine demonstrate, that one may stumble on such a rare find without any suspicion of it, while another pit may not have traces of posts indicating its function as a latrine. The latrine at St. Kolumba presented here was unquestionably not a singular construction in Roman Cologne.

(Translated by Stefanie Hoss)

Bibliography

Bouet, A. 2009. *Les latrines dans les provinces gauloises, germaniques et alpines.* (Gallia Supplement 59). Paris.

Catalogue Exhibition 2009. 'Le bain et le miroir. Soins du corps et cosmétiques de l'Antiquité à la Renaissance'. Musée Cluny, Paris.

Dodt, M. 2001. Römische Badeanlagen in Köln. *Kölner Jahrbücher* 34, 267–331.

Dodt, M. 2002. Römische Bauten in den nördlichen Insulae der Colonia Claudia Ara Agrippinensium. *Kölner Jahrbuch* 35: 571–698.

Dodt, M. 2003. *Die Thermen von Zülpich und die römischen Badeanlagen der Provinz Germania inferior* (online publication of PhD thesis University of Bonn see http://hss.ulb.uni-bonn.de/2003/0117/0117.pdf).

Dodt, M. 2005. Römische Bauten im südlichen Suburbium. *Kölner Jahrbücher* 38: 433–733.

Dodt, M. and Seiler, S. (forthcoming) Die Ausgrabungen an St. Kolumba in Köln.

Eibner, A. 1970. *Entwicklung und Werkstoffe der Wandmalerei vom Altertum bis zur Neuzeit*. Reprint edition 1926. Wiesbaden.

Hellenkemper, H. 1975. Architektur als Beitrag zur Geschichte der CCAA. In *Aufstieg und Niedergang der römischen Welt* II, 4: 783-824.

Hellenkemper, H. 1986. Wasserbedarf, Wasserverteilung und Entsorgung der CCAA. In K. Grewe (ed.), *Atlas der römischen Wasserleitungen nach Köln* (Rheinische Ausgrabungen 26): 193–214. Köln.

Hottentot, W. and van Lith, S. M. E., 2006. Römische Schönheitspflegemittel in Kugeln und Vögeln aus Glas. *Bulletin Antieke Beschaving* 81: 185–198.

Noll, W., Born, L. and Holm, R. 1972/1973. Chemie, Phsenbestand und Fertigungstechnik von Wandmalereien im römischen Köln. *Kölner Jahrbücher Vor- und Frühgeschichte* 13: 4-81.

Precht, G. 1971. Die Ausgrabungen um den Kölner Dom. Vorbericht über die Untersuchungen 1969/70. *Kölner Jahrbuch Vor- und Frühgeschichte* 12. Beilage 1.

Precht, G. 2002. Konstruktion und Aufbau sogenannter römischer Streifenhäuser am Beispiel von Köln (CCAA) und Xanten (CUT). In H. Kell, and G. Gogräfe, (eds), *Haus und Siedlung in den römischen Provinzen. Grabungsbefund, Architektur und Ausstattung*: 181-198. Homburg/Saar.

Puybaret, M.-P., Borgard, Ph. and Zérubia, R. 2008. Teindre comme à Pompéi: Approche expérimentale. In C. Alfaro and L. Karali (eds), *Vestidos, textiles y tintes. Estudios sobre la producción de biennes de consumo en la Antigüedad. Actas del II Symposium Internacional sobre textiles y tintes del Mediterráneo en el mundo antiguo* (Atenas, 24.-26. 11. 2005): 185–193. Valéncia.

Oelmann, J. 1923. Gallorömische Straßensiedlungen und Kleinhausbauten. *Bonner Jahrbücher* 128: 77-97.

von Petrikovits, H. 1977. Kleinstädte und nichtstädtische Siedlungen im Nordwesten des römischen Reiches. In H. Jahnkuhn, R. Schützeichel and F. Schwind, (eds), *Das Dorf der Eisenzeit und des Mittelalters*. (Abhandlungen Akademie der Wissenschaften Göttingen, Philosophisch-Historische Klasse, Folge 3, Nr. 101): 86-137.

Seiler, S. 1977. Ausgrabungen in der Kirche St. Kolumba in Köln, *Zeitschrift für Archäologie des Mittelalters* 5: 97–119.

Seiler, S. 1989. Die Kirche St. Kolumba in Köln und ihre romanischen Vorgängerbauten. In *Colonia Romanica. Jahrbuch Förderverein der romanischen Kirchen Kölns e. V.* IV: 146–157. Köln

Sommer, C. S. 1988. Kastellvicus und Kastell. Untersuchungen zum Zugmantel um Taunus und zu den Kastellvici in *Obergermanien und Raetien*. (Fundberichte Baden-Württemberg 13): 457-707.

Latrine pits in the Roman *vicus* of *Vitudurum* / Oberwintherthur (Switzerland)

Verena Jauch

Introduction

The *vicus* of *Vitudurum*, today Oberwinterthur, is situated in NE-Switzerland (FIgure 1) The settlement lay on an important main road, which ran from the legionary fortress of *Vindonissa* in the west eastwards towards Raetia. After the establishment of the settlement around Christ's birth, the large-scale construction of the western quarter 'Unteres Bühl' started around 7 AD. At least 16 plots had wooden rectangular houses with a common frontline and a common walkable *porticus*. The wooden houses were constantly altered and rebuilt until the 3rd century AD. The centre of the settlement lay at the 'Kirchhügel', which had been settled at the latest by 30 AD. After a fire around AD 70, the first stone buildings – a temple and a bathhouse – were erected there. In AD 294, the Kirchhügel was fortified with a wall with towers. An inscription names the instigator and the date as well as the name of the settlement '*Vitudurum*'. Finds and graves indicate that the settlement continued until the early medieval period. The results of the excavations have been published in the VITUDVRVM series (monographs of the Kantonsarchäologie Zürich vol. 1-10, for the *vicus* generally see Jauch 2002 and Hedinger and Jauch 2000).

Some of the pits discovered during excavations up until recent can be interpreted as latrines due to their filling. Two large-scale excavations are especially important here: One of them is the excavation at the western quarter 'Unteres Bühl' with about 3500 m² and the excavation 'Kastellweg' with about 2600 m². Both areas are distinguished by their very wet ground conditions and consequent excellent preservation of wood. The excavation 'Unteres Bühl' has been published as volume 6 of the monograph series of the Kantonarchäologie

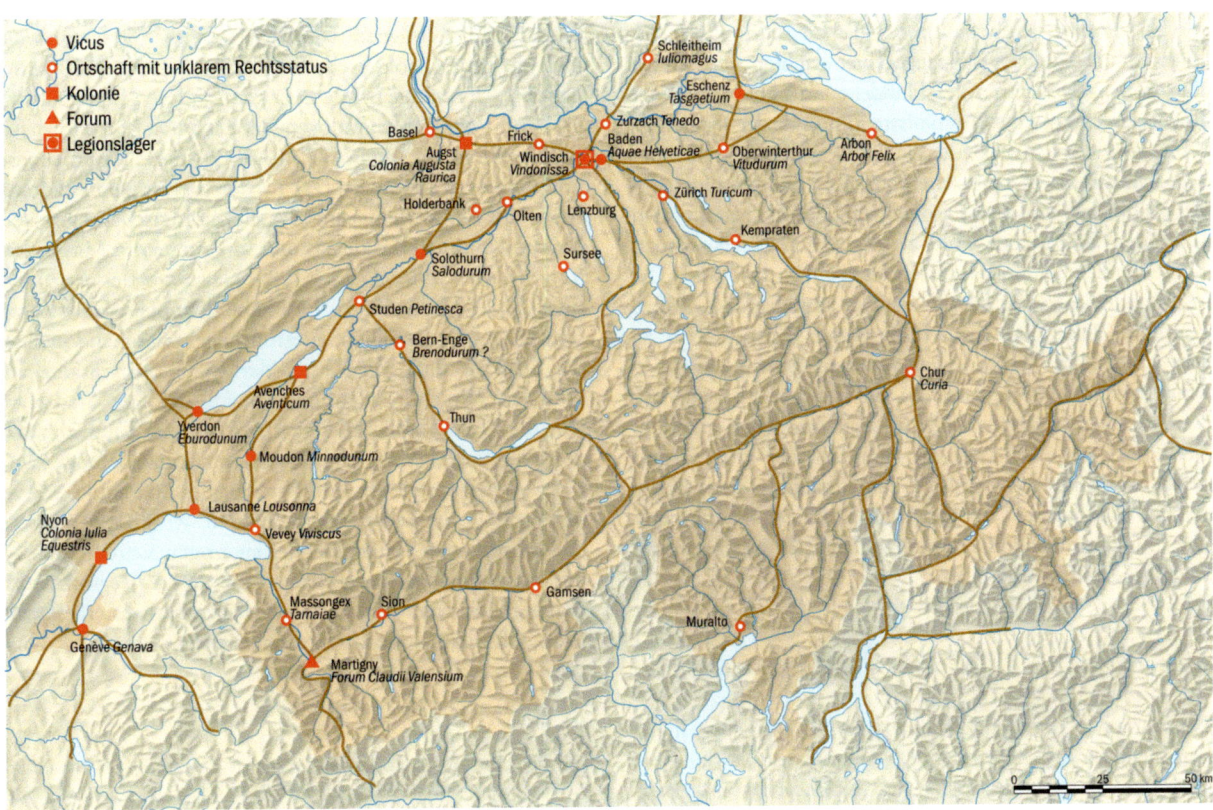

Figure 1. Switzerland around 15 BC – 20/30 AD. Cities of Celtic origin and Roman foundations (drawing S. Freudinger). Source: Flutsch L. *et al.* (ed.), Die Schweiz vom Paläolithikum bis zum frühen Mittelalter. SPM V. Römische Zeit (2002 Basel).

Zürich by Th. Pauli-Gabi, Ch. Ebnöther and P. Albertin (2002: 1-2); the excavation 'Kastellweg' has been finished in 2011 and is currently being analysed (Jauch and Zollinger 2010 (preliminary rapport); Jauch, Winet and Zollinger (in prep.).

The latrines

During the excavation 'Kastellweg', at least eleven latrines were discovered between 2005 and 2011. This was taken as an incentive to take a closer look at all latrines in the *vicus Vitudurum*. The overall number of latrines in the *vicus* is hard to determine, as the basis for interpretation often is insufficient. This is partly owed to the fact that some contexts had to be machine-dug. The main problem though, is the interpretation of the pits themselves. It is often difficult to determine which layers are from the period of use, as these are the only layers that offer clues to the function of the pit. In contrast, the final filling usually is made up of secondary refuse, which does not indicate the original function of the pit. As a rule, the latrines were situated in the back yards, several meters from the associated buildings. This is the reason why the layers of use usually do not correlate with the construction of new pits in the back yard. It is often difficult to determine, which floor level is associated with which pit and how they connect to the stratigraphy of the house.

The location of the latrines

1. The latrines are situated in a line in the backyard about 26-30 m or more behind the front of a residential building and form a uniform utilisation zone. The distance from the back of the houses is between 5 and maximally 15 m.

This seems to have been the standard location of the latrines in the *vicus*.

From the excavation 'Unteres Bühl' several examples of this type can be cited: In the back yard of plot 12, the pits were situated c. 15 m behind the backs of the houses (Figure 2; A9). A precise differentiation of the periods of the settlement horizon IV, dated between AD 70-90 and 120, was not possible, as the location could only be documented sketchily. The round pit 136 was lined with a sort of wooden weave (A9). Latrine 140 (A9) with a wooden lining and situated directly on the boundary of the plot, contained many fruit pits. The interpretation of the other plots is largely obscure. Seven pits on plot 18/20 (A11), situated between 10 to 13 m behind the assumed back of the house probably date to the same phase. No relative sequence could be reconstructed and no function could be ascertained. The picture is repeated in the first half of the 2nd century (Figure 2; A10, B1). The empty areas around the latrines are gardens. The access to the respective back yards was through the houses.

A uniform functional zone can be recognized in the neighbouring plots as well, where the pits lay c. 26-30 m from the front of the buildings.

Looking at the rest of the *vicus*, the picture for the first half of the 2nd century is quite uniform (Figure 3). The surface plan of the *vicus* clearly shows a division of all plots into the same functional zones. At the 'Kastellweg', the latrine belonging to a stone building of the 2nd century AD (G274; A15) was situated around 10 m from the back of the building, directly on the boundary to the neighbouring plot. The pit, categorized unmistakably by its contents as a latrine, had been abandoned during the 3rd century AD.

It was often possible to reconstruct from the context that the preferred location for the back yard latrines had been next to the boundary with the neighbouring plot. In the 'Kastellweg' excavation, pit 754 had been located right next to latrine 803 on the neighbouring plot during the first half of the 1st century AD. A fence marked the boundary between the two pits (Figure 4; A19; B3). Pit 803 (B3), which had been lined with wood in a manner reminiscent of a palisade, had been only about 2 m from the associated building. The wastewater drained to the south via an overflow; presumably there was an influx from the north. The drain ran along the western wall of a building and doubled as the eavesdrip drain for the roof water. Pit 754 (A19) was also lined in wood and probably belonged to a building about 10 m to the SW. We know of a similar arrangement from the 'Unteres Bühl' excavation, where pit 49 on plot 12 (A10) lay opposite pit 58 on plot 14 (B1) during the first half of the 2nd century (SH V, Figure 2). Both pits lay c. 9 resp. 13 m behind the back of their respective house.

2. The latrines lie right behind the house or at a close distance to the house, between 1,5 and 5 m; occasionally, they seem to have been housed in a sort of shed or outhouse.

A good example for this was found at Römerstrasse 186: Latrine 11 (A1) was situated about 5 m from the back of the house. Associated with it was a flushing water channel covered with boards. The installation dates to the period AD 40-55 to 80. Pit 14 (A2) was found in House C2, a square annexe to residential building C1. The organic filling with many pits and seeds indicates an interpretation as latrine. The allocation of the phases is not quite certain though, and was determined on the base of the finds alone. In a younger phase, latrine pit 15 was in an annexe (A3; Figure 5). The plan of house D2 was preserved in a rubble foundation (w. 1 m, d. 50 cm). J. Rychener reconstructs a stone building from this. The pit was constructed with four round posts and planks, which were fixed with diagonal posts. The peaty filling with again many pits and seeds allows an interpretation as latrine, while the finds date the filling into the mid-2nd century AD.

Figure 2. Unteres Bühl, pits in the back yards during the first half of the 2nd century AD (not to scale).

Figure 3. The *vicus* during the first half of the 2nd century with identified latrine pits (orange – not to scale).

Figure 4. Kastellweg, Two latrine pits of the first half of the 1st century AD. Left: pit 754, right: pit 803 (the dashed part shows the size of pit 803). A fence between the two marked the property boundary. (Photo: Ch. Lanthemann, Kantonsarchäologie Zürich).

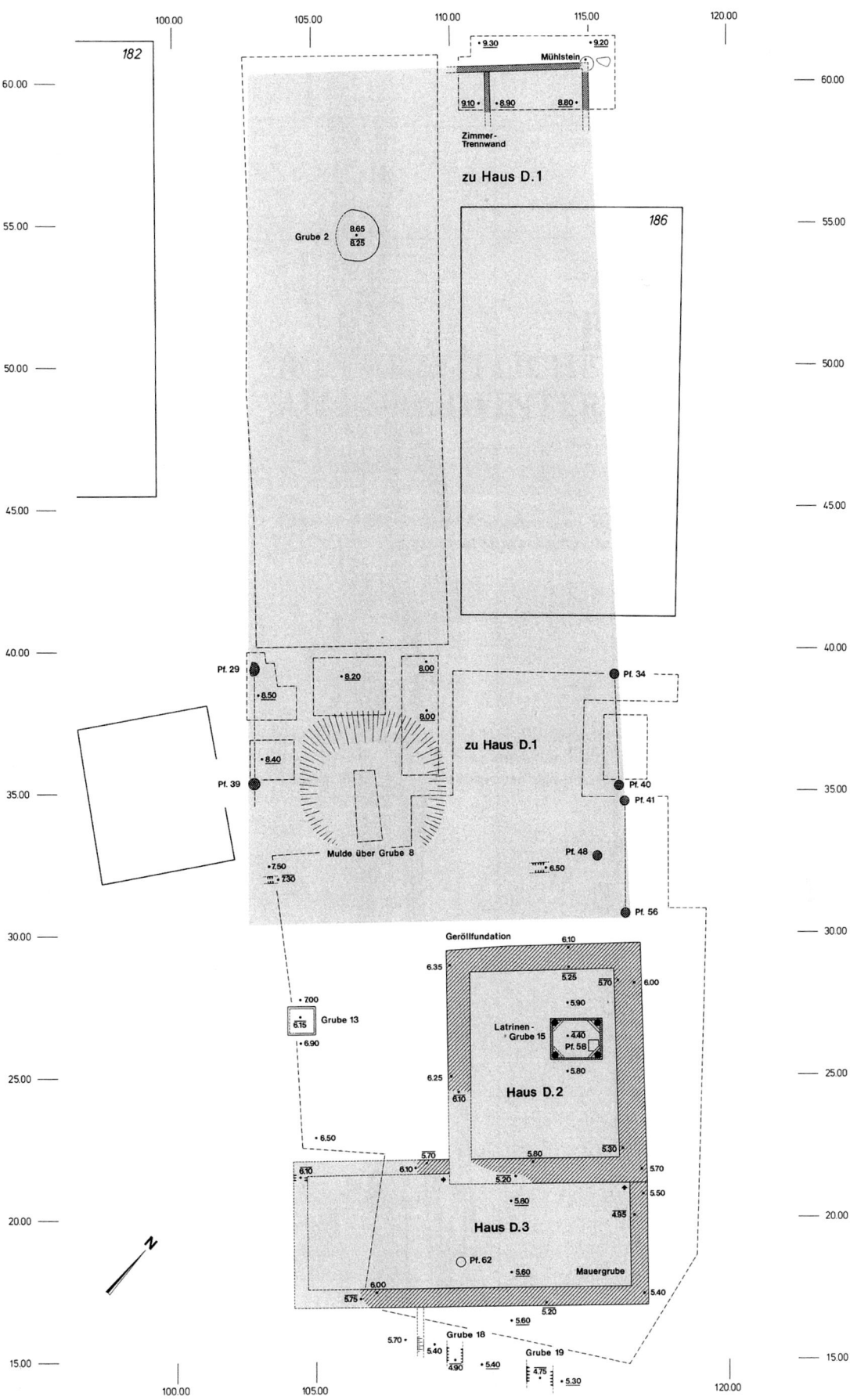

Figure 5. Römerstrasse 186, Rectangular wooden house with latrine pit no 15 in a an annex behind the house, 2nd century AD. Not to scale.

Figure 6. Kastellweg, latrine pit no 1470 with flushing system. The wooden step is recognizable on the northern side of the pit (Photo: B. Zollinger, Kantonsarchäologie Zürich).

The distance between residential building and latrine was small on some plots at the 'Kastellweg' excavation as well. One example is pit 1470, which lay just 1,5 m from the clay floor of the house (Figure 6; A20). At the same excavation, at least three other latrines were found in similar locations. Pit 790 (A16): 3,5 m behind the back of the house, filled in during the late 2nd century. Pit 762 (A17), 2nd century, probably belongs to a building 5 m to the south. Pit 803 (B3) (second quarter of the 1st century) lay just 2 m from the house.

3. The latrines are in a courtyard between houses. Lack of space leads to many pits overlapping.

The only example for this location is at plot Römerstrasse 227/229, where many pits were used as latrines – perhaps secondarily – in the second half of the 1st century (Figure 7, A5). According to their filling, the pits had been used as latrines at least secondarily (verifiable in pits 10, 11, 12, 14). The pits seem to be located in a courtyard that was c. 11 m wide (the southern expansion could not be measured), where they partly overlapped each other due to lack of space. A different interpretation of this context is that the pits lay in the back yard of a strip house oriented onto the main road and that the flanking buildings were workshops on neighbouring plots. Unfortunately, this area could not be further investigated archaeologically.

The design of the latrines

As already mentioned, the preservation of the wood was excellent in some places in the *vicus Vitudurum*. This fact allowed us to reconstruct the detailed design of the latrines – at least in part (see the catalogue of latrines below). As a rule, the latrines were rectangular or – less often – square. Very occasionally they had an oval form. The length of the sides were between 1 and 2 m, with a singular exception (Kastellweg, G274, A15) being 2,7 m long. The pits were conserved between 35 cm and 1,8 m deep. The preserved wooden linings consisted of four corner posts, behind which planks along the sides had generally been wedged. Occasionally, the planks had been placed in grooves in the posts, with some pits displaying a mix of both (Figure 8). Constructions made from vertically ramming boards or split branches or split trunks of young trees palisade-like into the pit were rare (Figure 9; see below B1-B4). Even rarer was the fitting of the pit with a basket: a large pit from the excavation Römerstrasse 155/157 (C1) had been lined

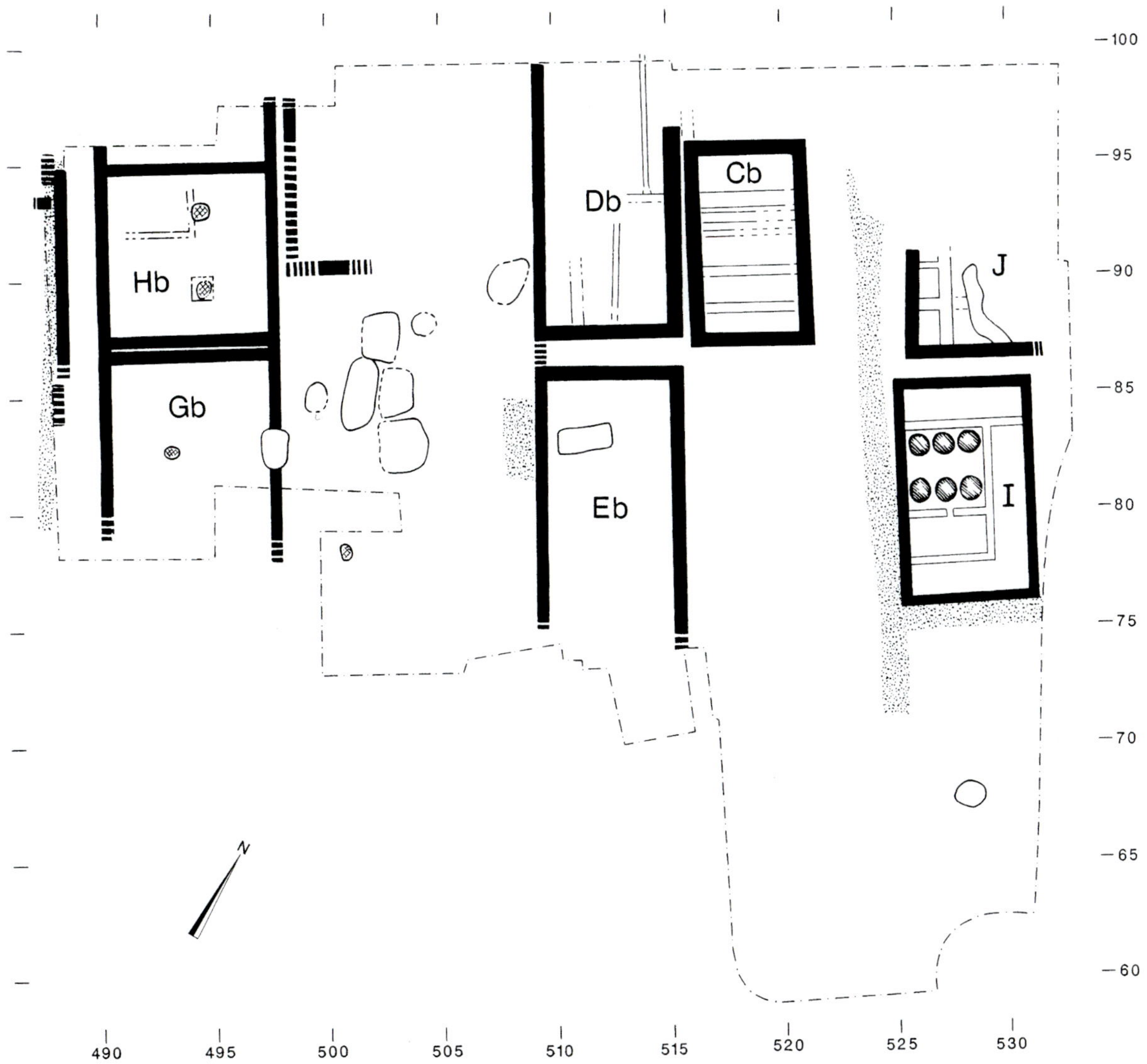

Figure 7. Römerstrasse 227/229, pits in the back yard between the houses, settlement horizont III, second half of the 1st century AD. Not to scale.

Figure 8. Pit encased in wood, corner posts with boards set horizontally in between.
Left: scheme, right: example Kastellweg, pit 754 (Photo: F. Mächler, Kantonsarchäologie Zürich).

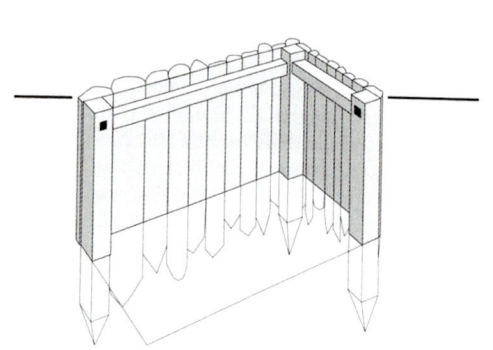

Figure 9. Pit encased in wood, boards rammed in vertically (like a palisade). Left: scheme, right: example Römerstrasse 177a, pit 575 (Photo: F. Jetzer, Kantonsarchäologie Zürich).

with two baskets placed into each other. A basket-weave installation is also mentioned for pit 136 (C5) from the 'Unteres Bühl' excavation.

Evidence for the seats

During the reconstruction, the question on the seats of the latrines was raised. Tangible evidence on the central placing of the opening had been found in pit 64 (Kastellweg, former Kirchweg, C5). In the centre of the pit, a discernible circular patch of faeces with a diameter of 40 cm was visible. The only pit with a cover made from wooden boards was pit 754 (Kastellweg, A19). As this cover has been preserved on just one half of the pit and as no seat was visible here, it seems likely that the seat had been on the half without a cover (Figure 10).

The roofs of the latrines

It can be assumed that latrines behind residential buildings were not set in the open air, but had at least a roof against the rain. The many shingle fragments found in the pits could have come from roofs, but it cannot be ruled out that they had been used to assuage the smell or even as a substitute for toilet paper.

Post-holes, whose posts may have belonged to either a roof or a seat were found in latrine pit G 332 of excavation Römerstrasse 169a (A13). This pit had been in use from the late 1st / early 2nd century and was abandoned towards the end of the 2nd century AD. The pit had the form of a parallelogram with side lengths of 1,8 x 1,9 m and a preserved depth of 1,4 m. Corner posts point to a disintegrated wooden lining. Two other posts (Pos. 727, 729) may be interpreted as possible remains of a roof construction. A similar construction was found for latrine pit 198 (Kastellweg; A21), where four postholes form proof for a roof. Nearby was another pit (G 274; A15), which had been in use during the 2nd century by the inhabitants of the building just 10 m off: both, the lowest organic filling and the two layers above that (which originated from the period of the abandonment of the pit), contained more than 100 kg roof tile fragments. The upper part of these probably originated from the demolition of the stone building associated with the latrine, while the tiles from the lower filling layers probably came from the roof of the latrine itself, including a large *tegula* with the decorative imprint of a dog's paw. The large amount of sizeable tile fragments found in pit 12 (Gebhartstrasse 18-22; A25) might also have been from a roof.

Latrine G575 (Römerstrasse 177a; B4) was situated at the NW-corner of an annexe and had two posts, indicating either a roof or an entrance. The toilet at Römerstrasse 186 (A2) was situated in an annexe of the building from the mid-1st to the mid-2nd century AD.

Toilet flushing

The latrine at Römerstrasse 186 was entered via a platform on the NW-side (A1). On the opposite side was an overflow, which led to a 40 cm wide and 35 cm deep channel (Figure 11). Traces indicate a cover with boards laid lengthwise. The channel had a gradient of 6,3 % towards the NE and fed into a drain (channel 21). This was a ground and surface water drainage leading along the eastern wall of the building and finally to the main street.

Figure 10. Kastellweg, pit 754, reconstruction (by Marcus Moser, Kantonsarchäologie Zürich). Not to scale.

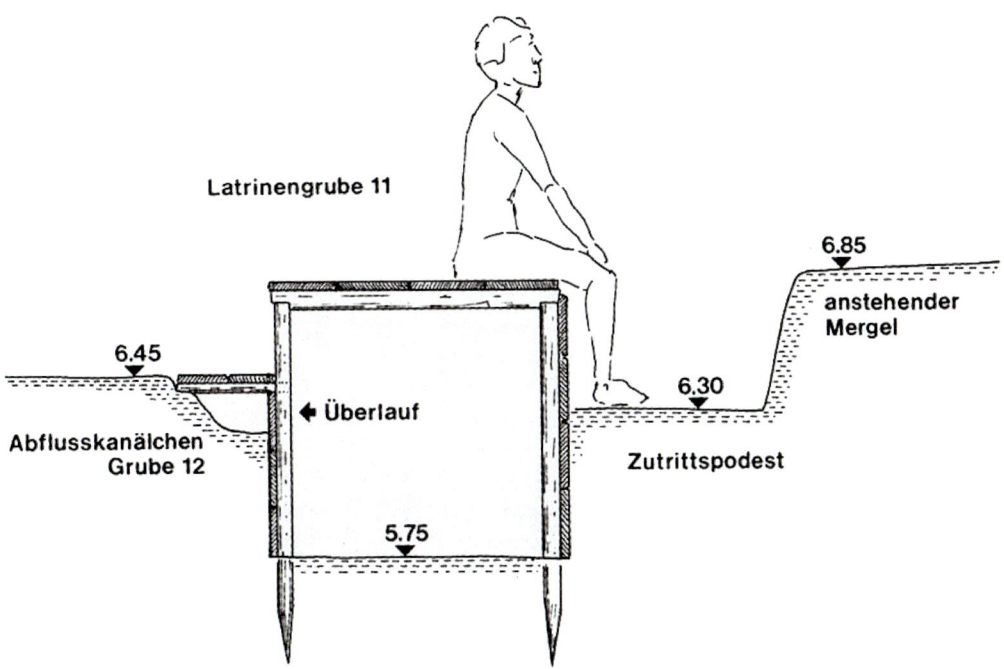

Figure 11. Römerstrasse 186, reconstruction of latrine pit 11. 40-55/80 AD.
(Rychener and Albertin 1986: 32, fig. 47).

Pit 64 (A24) lay at the rear boundary of the plot at the excavation Kastellweg, former Kirchweg. Along the eastern side ran a small ditch Pos. 89 (w. max. 26 cm, d. preserved max. 8 cm). This channel had a light gradient towards the latrine. The topmost plank of the wooden latrine pit lining had two 60 cm wide sections removed on opposite sides. This allowed the water to run into the pit on one side and out of it via the overflow on the opposite side. Pit 803 (Kastellweg; B3) also had a 30 cm wide overflow running into a small channel about 20 cm wide, made from half an oak tree trunk. The wastewater was transported southwards along the western wall of the building, with the drain doing double duty as an eavesdrip drain for the water from the roof.

At the excavation Püntenstrasse 2, a small drainage channel of 1,2 m length and 10 cm wide led away from pit 3 situated in the SE-corner of the back yard (A26). The installation probably dates to between AD 50 and 70.

Latrine G575 (Römerstrasse 177a; B4) displayed an impressively constructed flushing system: A plank had been placed along both the western and the northern side of the wood-lined pit, channelling the water coming from the north via a drain doubling as an eavesdrip drain for the water from the roof into the pit. Subsequently, the wastewater was evacuated via a rectangular channel on the building's southern wall. The pit was situated behind the building. Two posts indicate a roof or may have been part of the channel construction. Dendrochronology dates the installation to AD 58.

A possible latrine in Römerstrasse 169a (Pos. 296; A12) from the mid-Augustean / early Tiberian period (SH 1) had a small channel running into it which may have been a flushing system.

Access to the latrines

Most latrine pits were situated behind the residential buildings in the back yards and gardens. The access most likely was through the buildings on the front side of these long, rectangular plots. Signs of an entrance could be found in just a few installations: At Römerstrasse 186, a platform on the NW-side facilitated the entrance to the wood-lined pit (Figure 11; A1). This platform was about 55 cm above the bottom of the pit. On the opposite side lay the overflow to the drain.

In pit 1470 (Kastellweg), a wooden step made from oak was visible on the northern side (Figure 6; A20).

Spolia

The frequent re-use of the latrine timbers formed a continuous problem during the analysis of the contexts. The documentation of the excavations demonstrate that wooden elements left from the demolition of houses were recycled in the next building phase by being used in peripheral installations like latrines or drains. In extreme cases, the dates from the different timbers of one single installation can vary more than 100 years. An example for this is pit 49 (Unteres Bühl, A10), with the youngest corner post dating to AD 122 and two others to AD 121, fixing the date of the construction to after AD 122. The fourth post however, dates to AD 15.

In pit 58 on plot 14 (B1), a corner post with three grooves had been used, whose grooves remained unused, as the planks on the sides had been simply pushed down. A similar case is pit 754 (Kastellweg; A19), where the planks had also not been let into the grooves of the corner posts, which thus can be identified as *spolia*. Two planks of pit 790 (A16) date between 8 and 7 BC, the latter still with its outer layer! This is astonishing, as stratigraphy securely dates the pit to the 2nd century AD. From the four pointed corner posts of latrine 64 (Kastellweg, former Kirchweg; A24), one still carries an iron nail and another has a rectangular mortise in the lower part, obviously indicating it also was in secondary use in the pit. The reused corner posts were not analysed, two side planks – obviously not reused – which were cut in the autumn/winter 46/47 date the latrine pit.

Contents and interpretation

As a rule, the latrines were filled with large amounts of pottery, bones and other settlement refuse. In latrine 1470 (Kastellweg; A20), about 18 kg of pottery and 2 kg of brick were found. Because of the low grade of fragmentation of the finds, they probably originated in the residential building next to the latrine. In addition to that, easily visible fruit pits, fruit stones and seeds were found. In most fillings, pieces of shingles were discovered, probably the remains of the latrine roof. But they might also have had other functions, perhaps as covers for the malodorous filling or as a substitute for toilet paper (Bouet 2009: 171-172 enumerates several natural materials like straw, moss, plants or even stones). Several pointed sticks of about a finger thick were found in the primary fill of latrine 274 (A15), which, when fitted with moss or leafs, could have served the same purpose.

In some latrines, complete pottery vessels were discovered, most of them jugs, but both a beaker and a bottle have also been found, each in a different latrine (A15, A17, A20, A21, A25). All of the vessels were in all probability used for cleaning purposes and given up as lost once they had fallen into the latrine.

Some latrines have only a small amount of finds and accordingly must have not been used as refuse pits. An example for this is latrine 754 (Kastellweg; A19), which has been analysed archaeobotanically and

Figure 12. Kastellweg, a pair of (shoe) lasts made from maple wood, found in latrine 754 (Photo M. Bachmann, Kantonsarchäologie Zürich).

contained very little settlement refuse. Probably just before the abandonment of the latrine in the second quarter of the 1st century AD, a special object was placed under the cover (Figure 12; Jauch and Volken 2010, p. 221-240): In the SW-corner of the wooden box, at a depth of 30 cm, the excavator discovered a last for a left foot. About 18 cm under it lay a second last for a right foot, whose tip was stuck between the corner post and the sideboards. As shown in the reconstruction, the right last was deposited first, the left one following it (Figure 10). At the time of the deposition of the lasts, the latrine was almost filled. The reasons for the deposition of this pair of maple-wood lasts will probably remain unknown.

This text was finished in 2010. The current state of research on all latrines of Vitudurum excavated to date (including the ones mentioned here) will be presented in Jauch, Janke and Winet (in preparation).

Catalogue of the latrines in the Vitudurum vicus (as of 2009)

A. Rectangular and square pits, lined with horizontal boards

A1. RÖMERSTRASSE 186
Pit 11, building phase B and also C, dated AD 40-55/80 (Pit 20, building phase B und also C, AD 40-55/80 has been termed a latrine on plan 3 as well. But this interpretation is uncertain, as the pit had been mechanically excavated. However, wooden remains do indicate a wooden lining).
PLAN: 1,2 x 1,2 m, d. 0,7 m. 4 corner posts with sideboards pushed down, platform at NW-side as entrance. Opposite this, overflow into drain 12, with plank cover, w. 40 cm, d. 35 cm; gradient c. 6,4 % towards NE, led into open drain channel 21.
CONTENTS: many cherry stones.
LOCATION: about 4 m from a non-residential building (B2).
REFERENCES: Rychener and Albertin 1986: 32, plan 3.

A2. RÖMERSTRASSE 186
Pit 14, building phase C, dated second half 1st century AD.
PLAN: 1,6 x 1,1 m, d. 0,9 m.
CONTENTS: many fruit pits and seeds.
LOCATION: behind house C1, in annexe C2.
REFERENCE: Rychener and Albertin 1986: 41, plan 4.

A3. RÖMERSTRASSE 186
Pit 15, building phase D, dated 2nd century AD.
PLAN: 1,9 x 1,5 m, T. 1,5 m. Four round posts with sideboards pushed in, fixed with diagonal timbers.
WOOD: oak.
CONTENTS: many fruit pits, stones and seeds.
LOCATION: behind house D1, in annexe D2.

archaeozoologically. It contained a rich spectrum of food plants like grain, legumes, many spices, salad and vegetables, fruit and nuts.[1] Characteristic markers of faeces are concretion of small fragmented plant remains, testa fragments from grains, pit cells from pears and food plants with small seeds (millet, fig, wild strawberry and others). Larger fruit stones like plum stones and animal bones point to a function as refuse pit. A remarkable small amount of finds were discovered in the lowest layer of the filling: some pottery fragments, glass, a *tegula*, a white stone counter and a piece of iron plate. Pit 64 (Kastellweg, former Kirchweg; A24) also had almost no refuse in its contents: besides nine body sherds of coarse pottery and four bone fragments, only one iron nail was found.

Indications of a ritual function?

In one case, we may have encountered a ritual deposition in a latrine: as mentioned, pit 754 (Kastellweg; A19)

---

[1] Zoological analysis Heide Hüster Plogmann, IPNA, Basel, CH; botanical analysis Patricia Vandorpe, IPNA Basel, CH.

REFERENCE: Rychener and Albertin 1986: 44, fig. 72-73, plan 5.

A4. RÖMERSTRASSE 227/229
Pit G1, unknown date.
PLAN: 1,9 x 0,4 m, d. 0,4 m.
CONTENTS: many coprolites, primary function unknown.
LOCATION: in the remains of a house behind the residential building.
REFERENCE: Janke and Jauch 2001: 171, 274, fig. 202.

A5. RÖMERSTRASSE 227/229
Many pits of unknown function (e. g. pit 10, 11, 12, 14). Settlement horizon III, dated second half 1st century AD.
PLAN: round to rectangular. Pit 10 min. 1,5 m x min. 1 m, d. 0,6 m. Pit 12, plan: 1,5 x 1,2 m, d. 1 m.
CONTENTS: Because of content like shingles, seeds and plant remains used at least secondarily as latrine and refuse pit. Pit 10: the content is interpreted as faeces, mixed with stable manure.
LOCATION: in back yard between and probably also behind houses.
REFERENCES: Janke and Jauch 2001: 188, 210, 274-275, fig. 217; 222.

A6. RÖMERSTRASSE 227/229
Pit 18, settlement horizon IV, dated around AD 80-120, dendrodated AD 121 (with wane). Water collector or latrine. The latter is more likely because of the organic filling and the placing at the boundary of the plot.
PLAN: c. 1,5 x 1,5 m, d. 1,5 m. Wooden lining with corner posts and reinforced corners, set into the larger pit 19, which had been filled with clay.
WOOD: Oak.
CONTENTS: organic, partly sandy filling.
LOCATION: In the back yard on the border fence of a plot, no associated building excavated.
REFERENCES: Janke and Jauch 2001: 190-196, fig. 231. 240-243.

A7. RÖMERSTRASSE 227/229
Pit 20, settlement horizon IV, dated round AD 80-120, pit 20. PLAN: 1,5 x 1,5 m, d. 0,5 m.
CONTENTS: organic, wooden lining with corner posts. Certainly used as latrine in secondary use.
LOCATION: in back yard at boundary, associated building is not clear
REFERENCES: Janke and Jauch 2001: 190, 196, fig. 231. 243.

A8. UNTERES BÜHL, PLOT 9
Pit 5, dated AD 40-70, dendrodated AD 42. A function as central well is probable, was possibly only filled with refuse from the latrine or used as a latrine during secondary use.
PLAN: 1,75 x 1,5 m, d. 2,5 m.

WOOD: Oak.
CONTENTS: organic material, fruit pits, many wooden shingles.
LOCATION: behind the houses in the back yard or a central inner courtyard.
REFERENCES: Pauli-Gabi, Ebnöther and Albertin 2002/1: 88; Bd. 2, 251-253 fig. 450-460

A9. UNTERES BÜHL, PLOT 12
Settlement horizon IV, dated AD 70-90/120. Six pits in a row, functions unclear.
PLAN: Pit 140 (1,3 x 1,5 m), corner posts and pit walls made from horizontal split oak.
WOOD: Pit 140: Oak.
CONTENTS: Barrel pit 138 (diam. 1,1 m) contained a remarkable amount of fruit pits and stones.
LOCATION: Different pits in back yard, in a row about 15 m behind the back of the building. Badly documented.
CONTENTS: many fruit pits and stones.
REFERENCES: Pauli-Gabi, Ebnöther and Albertin 2002/2: 90-1. fig. 119, pits 136-140. esp. p. 98.

A10. UNTERES BÜHL, PLOT 14
Pit 49, Settlement horizon V, dated AD 90/120-130/150, dendrodated AD 121/122 (3 corner posts), one corner post dates to 15 AD.
PLAN: 1,8 x 1 m (outside), d. 1 m. Lined in wood, square oak posts and at least the planks and round timbers placed on top of each other.
WOOD: oak.
CONTENTS: Upper edge of bottom layer is organic layer.
LOCATION: in back yard opposite a second latrine on the neighbouring plot (G58 see part B1).
REFERENCES: Pauli-Gabi, Ebnöther and Albertin 2002/2: 120, fig. 175-176; Dendrodating: Labor für Dendrochronologie der Stadt Zürich.

A11. UNTERES BÜHL, PLOTS 18/19
Mid 1st to 2nd century AD. Seven pits in a row (G31-G33; G25-G38). Function unclear, a function as refuse pits is also possible.
PLAN: rectangular to square, with some wooden remains.
CONTENTS: organic to clayish.
LOCATION: 10-13 m behind the back of the house.
REFERENCES: Pauli-Gabi, Ebnöther and Albertin 2002/2: 151, fig. 239-240, esp. 163-164; Pauli-Gabi, Ebnöther and Albertin 2002/1: 138, fig. 116.

A12. RÖMERSTRASSE 169A
Pit 296, Settlement horizon 1, mid-Augustean to early Tiberian. Not completely excavated. Latrine or refuse pit.
PLAN: Lower edge diam. 2 m, Upper edge diam. c. 3,5 m, d. 70 cm. Water inflow from small channel Pos. 239 may be interpreted as flushing system.
CONTENTS: In the lowest layers, the coprolites, possibly of dogs, were found.

LOCATION: Back yard, associated building not excavated.
REFERENCES: Janke and Roth 2016: 38-39, Fig. 43-44.

A13. RÖMERSTRASSE 169A
Pit 332, settlement horizon 3-4, used from end of 1st / begin of 2nd century, abandonment end 2nd century.
PLAN: 1,8 x 1,9 m, T. 1,4 m. Flat bottom, vertical walls, marks of wooden lining (post Pos. 727 and 729).
CONTENTS: In the filling, a very thin grey clay at the bottom, above that a hard, brown deposition, excrements in pieces and pats above that, many wood shavings, shingles
LOCATION: Belongs to building B and sits just 1,2 m of its SE-corner.
REFERENCES: Janke and Roth 201: 52 Fig. 62; 57-58 Fig. 70-72.

A14. KASTELLWEG
Pit 419, dendrodated AD 26 (with wane)
PLAN: rectangular, >1,6 (rest in section) x 1,1 m, d. 40-47 cm; oak frame, one plank on the south side, on the western and northern side each two planks.
WOOD: Oak.
CONTENTS: organic, perhaps also manure.
LOCATION: back yard, behind residential building.
REFERENCES: Jauch V. in Jauch, Janke and Winet (in prep.); Labor für Dendrochronologie der Stadt Zürich, Kurt Wyprächtiger, Bericht Nr. 615.

A15. KASTELLWEG
Pit 274, dated second half of 2nd /1st half of 3nd century AD. On the bottom are remains of an older latrine (remnants of an oak construction).
PLAN: 1,5 x 2,7 m, d. 0,5 m. Round corner posts, 3 layers of planks, the lowest of which sat in grooves in the corner posts, while the upper were simply pushed down.
WOOD: alder.
CONTENTS: organic, many shingles, fruit pits and kernels, nuts. Two complete beakers.
LOCATION: around 10 m behind a stone building.
REFERENCES: Jauch V. in Jauch, Janke and Winet (in prep.).

A16. KASTELLWEG
Pit 790, dated beginning of 2nd century AD. Build out of spolia, H 217 dendrodated autumn/ winter 7 AD (with wane) and H 222 dendrodated 8 AD (no wane). Small drain, possibly associated.
PLAN: 1,3 x 1,2 m, d. 0,35 m. 3-4 layers of planks sat in grooves in corner posts.
WOOD: oak.
CONTENTS: organic.
LOCATION: 3,5 m behind the annex of a residential building.
REFERENCES: Jauch V. in Jauch,Winet and Zollinger (in prep.); Labor für Dendrochronologie der Stadt Zürich, Kurt Wyprächtiger, Bericht Nr. 654.

A17. KASTELLWEG, Pit 762
Dated last quarter 1st/first half 2nd century AD.
PLAN: l. 1,7 m, w.>0,7 m (rest in section), d. 0,55 m. No wood preservation.
CONTENTS: massive cherry pit horizon, whole bottle.
LOCATION: back yard, behind shed.

A18. KASTELLWEG
Pit 1402, dated second half of 2nd century AD. Several dendrodates of 159 AD.
PLAN: 2,1 x 1,6 m, d. 0,7 m. Wood lining, planks 1-4 layers. S-wall: 4; W and E-wall: each 3, N-wall at least 1 plank. The planks were pushed behind the corner posts.
WOOD: oak.
CONTENTS: organic, cherry pits, shingles.
LOCATION: ca. 15 m behind strip house.
REFERENCES: Gabriel A., Untersuchungen der Latrinengrube Pos. 1402, Kastellweg Parz. 3194. Seminararbeit Universität Zürich, Abt. für Ur- und Frühgeschichte (2010); Unterwasserarchäologie und Dendrochronologie (UWAD) der Stadt Zürich, Niels Bleicher and Karolina Widla, Bericht 945.

A19. KASTELLWEG
Pit 754, dated second quarter of the 1st century AD. Wood of cover (with wane) dendrodated AD 32.
PLAN: 1,4 x 1,8 m, d. max. 0,6 m. Wooden lining, four grooved corner posts. The planks were not let into the grooves of the corner posts and thus must be *spoila*. Beechen cover in southern half, fixed with oak crossbar.
WOOD: Corner posts and 4 planks oak, 3 planks and cover beech, 3 planks poplar.
LOCATION: ca. 7 m behind building.
REFERENCES: Jauch and Volken (2010); Jauch V. in Jauch, Janke and Winet (in prep.); beech nr. 240: oral communication Niels Bleicher, Labor für Dendrochronologie der Stadt Zürich.

A20. KASTELLWEG
Pit 1470, dendrodated (with outer layer) AD 78 (with wane), fitting the dates of the archaeological finds.
PLAN: 1,3 x 1,3 m, d. 0,6 m. Wooden lining, corner posts have been hewn square, planks set in grooves. Walls made up of 3 to 4 planks. Wooden step on northern side.
WOOD: oak.
CONTENTS: organic, mineralised plant remains. Complete jug, bronze bell.
LOCATION: 1,5 m behind a residential building.
REFERENCES: Zellwegger J., Zürcher Ch., Latrine 1470 – Kastellweg, Oberwinterthur. Bachelorarbeit an der Universität Zürich, Abteilung für Ur- und Frühgeschichte (2010); Jauch V. in Jauch, Janke and Winet (in prep.); Labor für Dendrochronologie der Stadt Zürich, Niels Bleicher, Bericht Nr. 691.

A21. KASTELLWEG
Pit 198, dated second half of 2nd/first half of 3nd century AD.

PLAN: 2,2 x 1,8 m, d. min. 1 m. Wooden lining is missing, only one corner post preserved. Four post-holes belonged to a roof construction.
CONTENTS: organic.
LOCATION: about 12 m behind a residential building. Complete beaker.
REFERENCES: Jauch V. in Jauch, Janke and Winet (in prep.).

A22. KASTELLWEG
Pit 739, dated last quarter 1st/beginning 2nd century.
PLAN: >1,7 x >0,8 m, d. >20 cm. Wooden lining is missing.
CONTENTS: organic.
LOCATION: 1 m behind a residential building.
REFERENCES: Jauch, Janke and Winet (in prep.)

A23. KASTELLWEG
Pit 1276, dated 1st quarter 1st century.
PLAN: 1,6 x 1,35 m, d. >45 cm. Wooden lining is missing.
CONTENTS: Organic filling, many small pieces of wood (analysed by Patricia Vandorpe und Simone Häberle, Institut für prähistorische und naturwissenschaftliche Archäologie, Universität Basel, IPNA) point towards function as latrine.
LOCATION: back yard, relation to building unclear.
REFERENCES: Jauch, Janke and Winet (in prep.)

A24. KASTELLWEG, FORMER KIRCHWEG
Pit 64, dated 1st century AD. Youngest dendrodate of timber No. 23 (with wane) spring AD 47 and timber No. 27 (with wane) autumn AD 46. Traces of timber working on the corner posts point towards their secondary use.
PLAN: 1,7 x 1,4 m, d. max. 0,5 m. Lined with wood, corner posts with planks set in grooves, S-wall 2 planks, the other walls 3 planks each. 4 pointed oak posts, one of them with a nail. On the eastern side secured with small posts. Parallel to the basin plank forming the side of a channel at a distance of ca. 20 cm. The water in the channel probably ran through an inflow cut into the top plank of the basin into the pit. Into the topmost preserved plank (No. 25), a cut-out had been made, so that an inflow area of about 60 cm was created.
WOOD: Oak.
CONTENTS: organic, many fruit pits, kernels, walnuts. In the centre of the bottom a circular hard patch of faeces with a diameter of 40 cm.
LOCATION: about 6 m behind shed in back yard.
REFERENCES: Jauch, Janke and Winet (in prep.); Labor für Dendrochronologie der Stadt Zürich, Kurt Wyprächtiger, Bericht Nr. 538.

A25. GEBHARTSTRASSE
Pit 12, dated to 2nd century AD (excavation unpublished). Many large brick fragments probably indicate former roofing, but no posts recognizable.
PLAN: Upper part circular, diam. 2 m, lower part rectangular 0,65 x 1,4 m, d. 1,8 m. With wooden lining, but without recognizable construction details. At the bottom clayish deposition, probably occasioned by water.
CONTENTS: Peaty, humus-organic. At the bottom completely preserved jug with one handle.
LOCATION: probably in back yard behind residential buildings.

A26. PÜNTENSTRASSE 2
Pit 3, dating not possible through stratigraphy, probably building phase B1, dated c. AD 50-70.
PLAN: 0,95 m x 0,7 m, d. 0,7 m. From SE-corner small drainage channel, l. 1,2 m, w 10 cm.
LOCATION: probably behind a strip house.
REFERENCES: Rychener 1988: 75, fig. 511.

A27. RÖMERSTRASSE 177A
Pit G465, ated second half of 2nd/beginning 3rd century AD.
PLAN: 2 x 2 m; d. 0,5 m. Remains of a wooden lining.
BOTTOM FILLING: organic, wood chips.
LOCATION: on gravel area in back yard.
REFERENCES: Jauch, Janke and Winet (in prep.)

B. Rectangular or square pits with vertical (palisade-like) wooden lining

B1. UNTERES BÜHL, PLOT 14
Pit 58, settlement horizon V, pottery dated AD 90-130.
PLAN: 1,55 x 1,1 m (outside), d. c. 1,2 m. Palisade-like placed posts next to each other. Posts made from split trunks of young trees, round timber and roughly shaped planks. In the corners oak posts, between which a stay was positioned on each of the four sides to secure the post walls. The corner post No. 1192 has grooves on three sides and must be a *spolia*, the other three corner posts are split trunks.
WOOD: 5 x juniper, the rest oak.
CONTENTS: many fruit pits and stones and wood chips.
LOCATION: in back yard opposite a second latrine on the neighbouring plot (G49 see part A10).
REFERENCES: Pauli-Gabi, Ebnöther and Albertin 2002/2: 120-1. fig. 177-178.

B2. UNTERES BÜHL, PLOT 12
Pit 139, settlement horizon IV, dated AD 70-90/120. Pit 139 from a row of six pits (A9), function unknown.
PLAN: 1 x 1 m. Wall made of standing split rails, with shakes in between.
WOOD: Beech and oak.
LOCATION: Different pits in back yard, in a row about 15 m behind the back of the building. Badly documented.
REFERENCES: Pauli-Gabi, Ebnöther and Albertin 2002/2: 90-1. fig. 119, pits 136-140. esp. p. 98.

B3. KASTELLWEG
Pit 803, dated to second quarter of the 1st century AD. Wood No. 184: 19 AD, No. 185: 19 AD with wane autumn/winter.

PLAN: 1,7 x 1,7 m, d. c. 1 m. Palisade-like wooden lining. Flushing system by overflow in channel Pos. 831. The channel runs along the western wall of the associated residential building.
WOOD: split trunks of young oak.
CONTENTS: Organic, many fruit pits, kernels and small pieces of wood.
LOCATION: directly behind a building, on the fence to the neighboring plot.
REFERENCES: Labor für Dendrochronologie der Stadt Zürich, Kurt Wyprächtinger, Bericht Nr. 654; Juach, Janke and Winet (in prep).

B4. RÖMERSTRASSE 177A
Pit G575 with Flushing system, building phase 3, dated to AD 50/60-100, finds dating to around AD 50. Dendrodated AD 58 (with wane).
PLAN: 1,3 x 1,3 m, d: 1,4 m., palisade-like wooden lining. Along the W and N side of the pit is a plank each, which channels the water led in from an eavesdrip drain of roof water in the north into the pit. The wastewater could be subsequently drained by a rectangular channel along the buildings' southern wall. The pit lay behind the building. Two posts indicate a roof or an entrance.
WOOD: Oak.
CONTENTS: organic material, faeces, shingles.
REFERENCES: Janke R. and Winet J. (in prep.); Bericht Kurt Wyprächtiger, Dendrolabor des Büros für Archäologie der Stadt Zürich, LEERSCHLAG 16.2.1996 (Archiv KA).

C. Oval pits

C1. RÖMERSTRASSE 155/157
Pit 126, dated second half of 2nd/3rd century AD. Interpretation uncertain.
PLAN: in upper part 2,3 x 3,4 m, d. 1,8 m. 2 baskets set into each other, only a quarter of the pit was excavated. At the edge of the basket-weave, the pit becomes rectangular: l. 2 x 1,4 m, bottom flat, d. 1, 6 m.
LOCATION: in back yard.
REFERENCES: Roth R., Pit 126 – Römerstrasse 155/157. Auswertung des Fundmaterials. Bachelorarbeit an der Universität Zürich, Abteilung für Ur- und Frühgeschichte (2012).

C2. RÖMERSTRASSE 227/229
Pit 14, settlement horizon III, dated second half of 1st century AD. Interpretation: refuse pit or latrine.
PLAN: long oval, l. 3,4 m, w 1,3 m d. 0,5 m.
CONTENTS: organic.
LOCATION: in courtyard or probably behind strip house.
REFERENCES: Janke and Jauch 2001: 275, fig. 222.

C3. RÖMERSTRASSE 227/229
Pit 11, settlement horizon III, dated second half of 1st century AD. Function uncertain.
PLAN: circular, diam. 1,3 m, d. 1 m.

CONTENTS: many small pieces of wood and macro remains, indicating a use as refuse pit or latrine.
LOCATION: in back yard or probably behind the strip house.
REFERENCES: Janke and Jauch 2001: 274-275, fig. 222.

C4. RÖMERSTRASSE 227/229
Pit 21, settlement horizon IV, dated AD 80-150.
PLAN: round-oval diam. upper part 1,9 m; diam. lower part 1,2 m; d. 2,3 m, was interpreted by R. Janke as well or latrine.
LOCATION: uncertain.
REFERENCES: Janke and Jauch 2001: 195, fig. 231.

C5. UNTERES BÜHL
Pit 136, dated AD 70/90-120
PLAN: circular, diam. 0.9 m, clad with wickerwork.
LOCATION: Different pits in back yard, in a row about 15 m behind the back of the building. Badly documented.
REFERENCES: Pauli-Gabi, Ebnöther and Albertin 2002/2: p. 90, fig. 119, p. 98.

(Translated by Stefanie Hoss)

Bibliography

Bouet A. 2009. *Les latrines dans les provinces gauloises, germaniques et alpines* (Gallia 59 Supplement).

Hedinger, B. and Jauch, V. 2000. Die römische Zeit. In: Hintergrund-Untergrund. Archäologische Entdeckungsreise durch Winterthur. *Neujahrsblatt Stadtbibliothek Winterthur* Band 331: 47–75.

Janke R. and Jauch V. 2001. *Beiträge zum römischen Oberwinterthur - Vitudurum 9. Ausgrabungen auf dem Kirchhügel und im Nordosten des Vicus 1988-1998.* (Monographie der Kantonsarchäologie Zürich 35). Zürich/Egg.

Janke R. and Roth M. 2016. *Forschungen im Zentrum des Vicus Vitudurum. Ausgrabungen an der Römerstrasse 169a und 173 sowie andere kleine Untersuchungen* (Zürcher Archäologie 33) Zürich/Egg.

Jauch, V. 2002. Winterthur Z H. Oberwinterthur. In: L. Flutsch, U. Niffeler and F. Rossi (eds), *Die römische Epoche* (Die Schweiz vom Paläolithikum bis zum Mittelalter Band 5): 403-404. Basel.

Jauch V., Janke R. and Winet I. (in preparation). *Baubefunde im Nordquartier des römischen Vicus Vitudurum, Oberwinterthur (Schweiz)* (Monographien der Kantonsarchäologie Zürich. Beiträge zum römischen Oberwinterthur 2 vols)

Jauch V. and Volken M. 2010. Ein Paar römische Schuhleisten aus dem *vicus Vitudurum* – Oberwinterthur (Schweiz). *Germania* 88: 1-2, 221-240.

Jauch V. and Zollinger B. 2010. Holz aus *Vitudurum* – Neue Entdeckungen in Oberwinterthur. *Archäologie Schweiz* 33: 2-13.

Pauli-Gabi Th., Ebnöther Ch. and Albertin P. 2002. 1-2, *Vitudurum 6. Ausgrabungen im Unteren Bühl.* (Monographie der Kantonsarchäologie Zürich 34, Bd. 1 und 2). Zürich/Egg.

Rychener J. and Albertin P. 1986. *Ein Haus im Vicus Vitudurum - Die Ausgrabungen an der Römerstrasse 186.* Beiträge zum römischen Oberwinterthur - Vitudurum 2. (Berichte der Zürcher Denkmalpflege, Monographien 2) Zürich.

Rychener J. 1988. *Beiträge zum römischen Oberwinterthur - Vitudurum* 3 (Berichte der Zürcher Denkmalpflege, Monographien 6). Zürich.

A Roman cesspit from the mid-2nd century with lead price tags in the civil town of Carnuntum (Schloss Petronell/Austria)

Beatrix Petznek

The following article is an introduction to the recent find of a cesspit with an extraordinary fill, whose full analysis was fortunately made possible by a generous grant from the Anniversary Fund of the Austrian National Bank (Oesterreichische Nationalbank, Jubilaeumsfund, Projektnr. 15007). (Figures 1 and 2).

Excavation

During the renovation of the baroque Schloss (castle) of Petronell in the municipality of Petronell-Carnuntum (40 km east of Vienna, Austria) in the year 2011, a latrine cesspit was discovered and subsequently excavated and documented by the ARDIG Archaeological Services Company (http://www.ardig.at). The cesspit is situated in the south, directly next to the south-eastern tower of the castle. It is surrounded by modern water pipes, drains and cesspits and it is something of a miracle that it had remained completely preserved until the excavation in spite of the building work for the nearby Schloss in both the renaissance and baroque periods (Figure 3).

Somewhere during a period around the mid-2nd century AD, a square shaft of 2,10 m width and 3,80 m depth was dug into the natural gravel-deposit in the northeast of what was then the civil town of *Carnuntum* in the province of Upper Pannonia. This was secured on the inside by wooden posts and boards, of which only some discoloration on the walls and the imprints of the posts in the corners remained on excavation (Fig 4).

On top of this underground construction, a wooden shed was erected both for privacy and as protection against the weather (Figure 5).

The cesspit was only used for a short time, but numerous thin chalk layers prove the intensity of the use. The upper part of the cesspit was disturbed by a secondary pit, which was created by partly emptying

Figure 1. Schloss Petronell, bird's eye view (Photo: R. Thoma ARDIG)

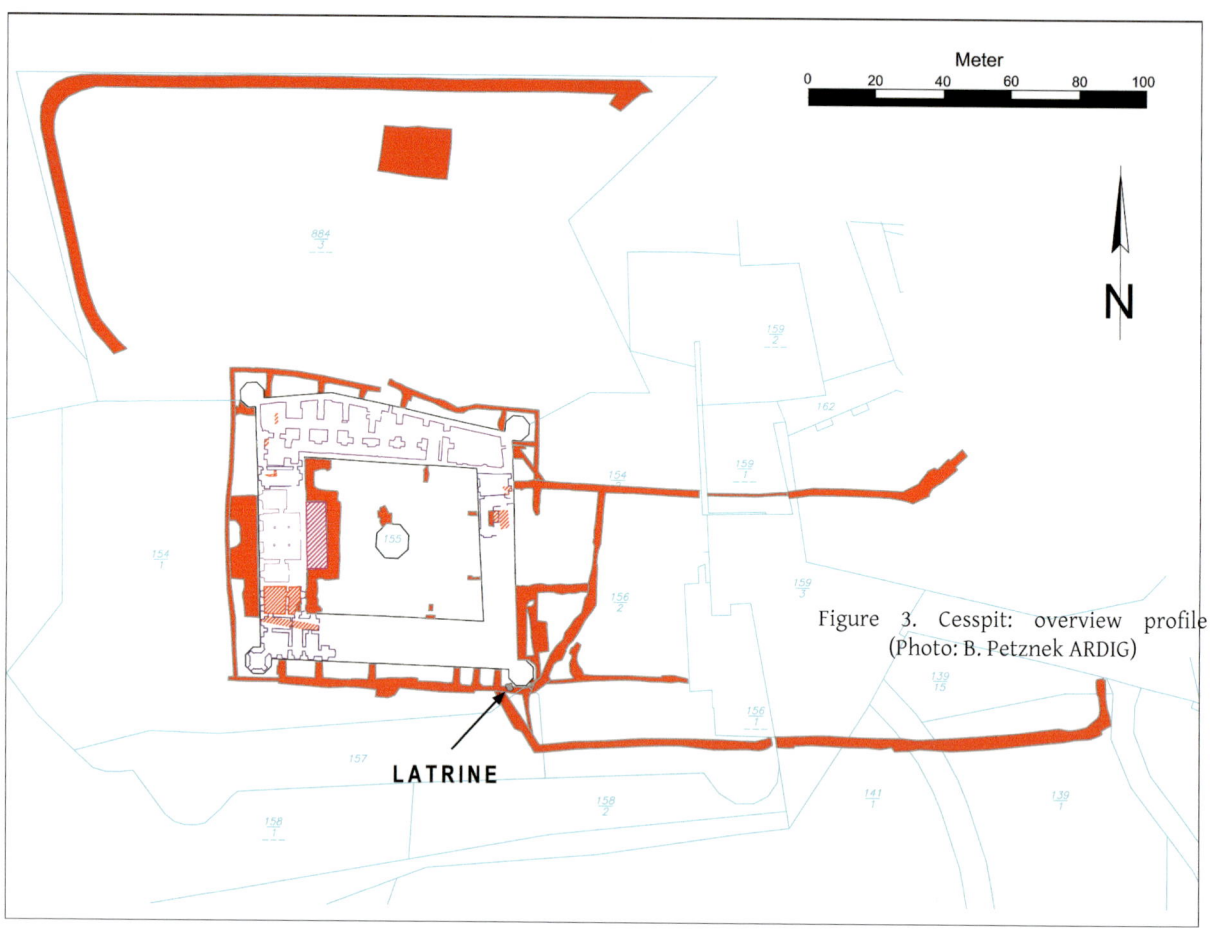

Figure 2. Schloss Petronell: plan of excavation and cadastre (© ARDIG after M. Raab, B. Petznek, Grabung Schloss Petronell).

Figure 3. Cesspit: overview profile (Photo: B. Petznek ARDIG)

Figure 4. Cesspit: interface (Photo: B. Petznek ARDIG)

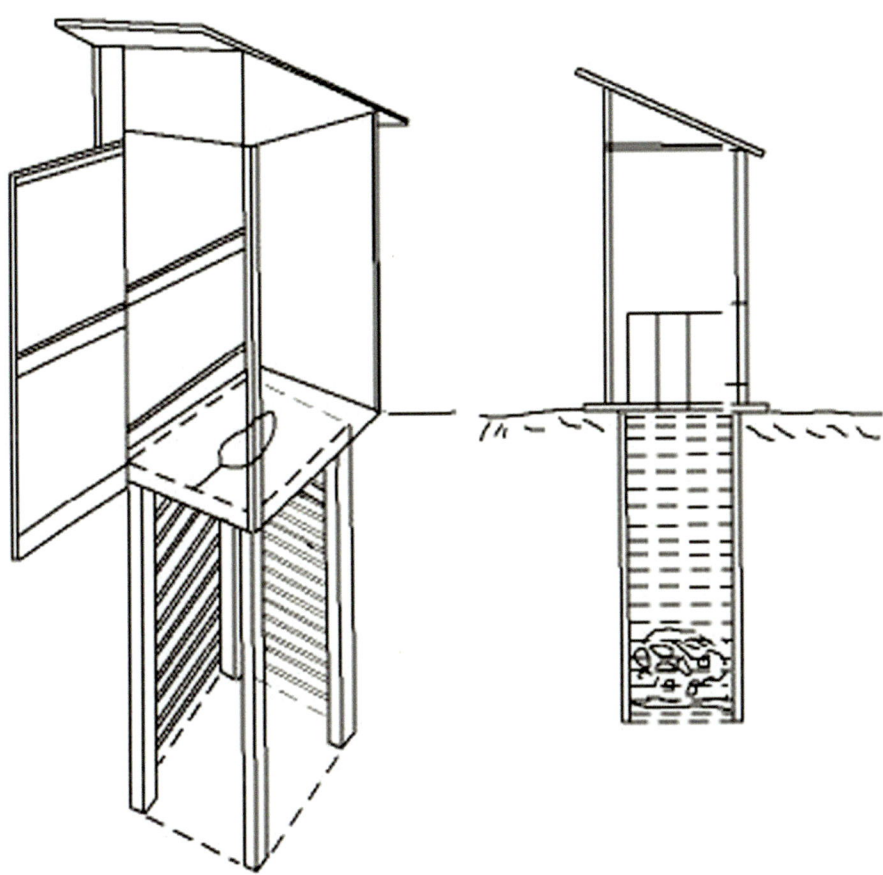

Figure 5. 3D-reconstruction of the cesspit (Drawing: B. Petznek ARDIG)

Figure 6. Profile of the cesspit (Photo: B. Petznek ARDIG)

the cesspit. The fill of this pit consisted of larger pieces than the lower part, with some large stones, tiles and roofing tiles. After the cesspit had been abandoned, the secondary pit was dug and filled and covered with larger pieces of screed (under-flooring), mosaic floors and other building material (Figures 6 and 7).

The undisturbed lower part of the cesspit, however, was filled with faeces and household waste. The refuse in the latrines can be categorised into two groups, namely items that were lost unintentionally and those that were thrown away on purpose. Coins, beads, knives, dice, gaming pieces and bone hair pins belong to the first group, while kitchen refuse, sweepings and the scraps of meals must have belonged to the second group. These consisted of animal bones, the shells of mussels and snails, broken vessels, building materials and workshop waste (see below).

Pottery

The pottery includes the usual ceramic forms of the first half of the 2nd century AD. Among the common wares are dishes, lids, jugs and pots with the typical rim forms (triangular and horizontal) of this period. Interesting is a fine cup with cornice rim, in whose bottom the letters 'CARN' were carved. (analysed by Beatrix Petznek, ARDIG)

The Samian Ware, analysed by Stefan Groh (OeAI), show a very homogeneous picture of the production sites. Imports from the pottery centres La Graufesenque and Banassac (Southern Gaul) and Lezoux (Middle Gaul) are almost equal in mass. A fragment of Arretina (presumably already old when lost) and two early fragments from Rheinzabern and one from Blickweiler complete the spectrum. The chronological range of the relief-decorated and stamped Samian Ware is fairly wide, from 100 – 170/180 AD. The main portion, however, dates into the period between 100 and 160 AD, only the fragments of bowls form Dragendorff 37 by CINNAMVS and IANV(ARIVS) continue to the end of the 170s AD.

The dating spectrum of the Samian ware, the common wares and the small finds as well as the coin dates (provided by Ursula Schachinger) all point towards to the sixth decade of the 2nd century AD as the time when the latrine was filled and the lead tags were abandoned (Groh and Petznek 2012: 68).

Lead commercial tags

In addition to these fairly common finds, a large lead plate (47 x 65 cm) and 240 inscribed rectangular pieces of lead sheeting were found in the fill. These are small and almost square (c. 3 x 4 cm) and were used as price tags. Price tags usually are small pieces of lead sheeting

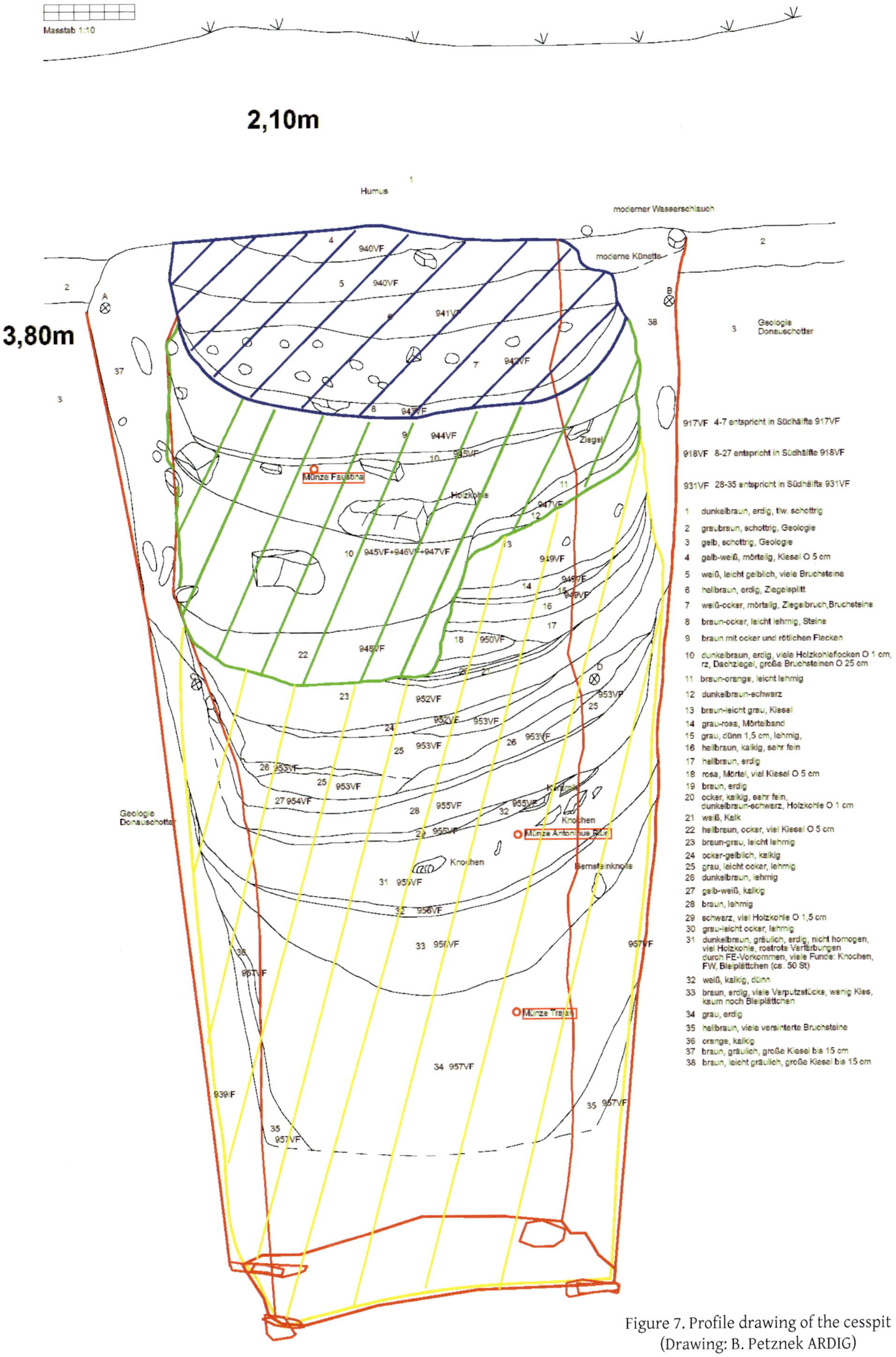

Figure 7. Profile drawing of the cesspit
(Drawing: B. Petznek ARDIG)

inscribed with at least the name of the product offered and its price.

The fronts and backsides of these tags have been inscribed with letters in Latin cursive writing c. 2 to 4 mm in size. The texts are brief – three to four lines – and consist of personal names (owner) and some short information on the commodities (clothes) or services (fulling, dyeing) offered plus their prices.

This find is exceptional as it is the first in the whole of the Roman Empire with such a large number of well-preserved leaden labels found in one closed find complex, which also is well dated by coins and ceramics. The information on the labels will allow researchers to gain an insight into which goods and services were offered at which prices in *Carnuntum* during the mid-second century.

A comparison with another large find with over 1000 (actually 1123) price tags from Siscia, also in Upper Pannonia (Sisak/Croatia, see Radman 2014), dating in the 1st and early 2nd century AD may even help to get an idea of the economic development and inflation in Upper Pannonia.

The find of these price tags will allow statements about the local economy in *Carnuntum* and also enable comparisons to other economic areas of the Roman Empire. In addition to this, palaeographic and onomastic studies will hopefully reveal information on how many writers wrote the labels and their names, origins, social position and profession.

All lead labels have been X-rayed to ensure they were inscribed and the images digitalized and optimized with a computer program. Because the milieu of the cesspit was rich in chalk and phosphate, the surface of the lead price tags has been affected and the inscriptions are now very shallow. Some price tags also have erased parts or ligatures, which further complicates the reading. To ensure correct reading, all tags were scanned in a computer tomograph and all images, in addition to the originals were sent to Ivan Radman-Livaja (Archaeological Museum Zagreb, Croatia), who is studying them. The examples already deciphered all name different items of clothing and their different colours (Figure 8a-e).

The name in the first two lines of the example pictured in Figure 6 most likely belongs to a peregrine woman, Verrucila Magiri. The name Verrucila seems to be a hapax, but names with the same root, such as Verrucius or Verrucosus are known in Noricum and Pannonia. The name of her father, Magirus is known from Northern Italy and is likely of Celtic origin.

After they had been read, twenty of the price tags and the large lead plate were sampled to provide material for lead isotope testing, financed by the Austrian Archaeological Institute (ÖAI) and carried out by Roland Schwab (Curt-Engelhorn-Zentrum Archaeometry, Reiss-Engelhorn-Museen Mannheim) and Ernst Pernicka (Institute for Archaeometry and Archaeometallurgy, University Heidelberg).

This method can show the origin of the lead, which will reveal information on the Roman trade in lead. It was formerly assumed that all Roman lead came from either England or Spain, but recent research has proven that smaller lead deposits in Germany and elsewhere were also exploited. The results could show that the lead for the labels probably came from Dalmatia, Dacia, southern Pannonia or Moesia Superior.

The lead price tags name the Latin names of colours, which could be correlated by Regina Hofmann-de Keijzer (Department of Archaeometry, University of Applied Arts Vienna), Andreas G. Heiss (OeAI) and Beatrix Petznek to the colour agents respectively dye plants as their possible origins as well as to the written sources.

Zoological and botanical remains

The careful recovery and conservation of the archaeological sediments during the excavation permitted the methodical examination of the zoological and botanical remains and the parasites.

The animal remains (bones, shells, insect remains) were examined by Alfred Galik (OeAI and Institute for Anatomy, Histology and Embryology, Vetmeduni Vienna). These demonstrate a rich diversity: Insect remains prove the presence of woodlice, ants, bees and flies, but also the fly larvae and pupae typical for latrines. In addition to insects, four-footed vermin like rats and mice also were found.

Chicken consumption can be proven through eggshells and bones, other birds consumed were ducks, geese, bitterns, partridge and small songbirds. Fish are represented by numerous species like river barbel, carp, catfish, perch, pike, roach and trout. Moreover, five mackerel vertebrae illustrate the import of marine fish, as does the single oyster shell. Frequent fragments of edible snails can be interpreted as consumption refuse, but the most important wild animal for consumption seems to have been the hare. Among the domesticated animals cattle dominate, followed by pigs.

Nearly 500 remains of plants could be determined by Ursula Thanheiser (VIAS Vienna Institute for Archaeological Science, University of Vienna) and Andreas G. Heiss (OeAI, formerly VIAS), representing 69 different species. The identified plant spectrum represents a wide variety of food plants, ranging from

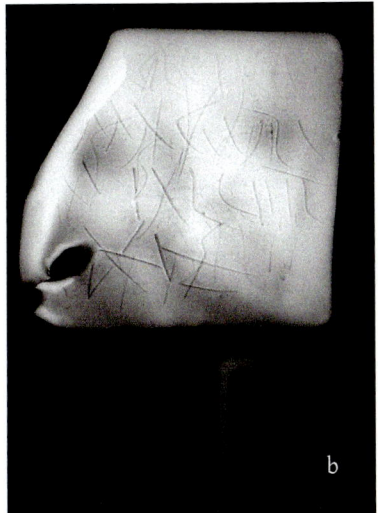
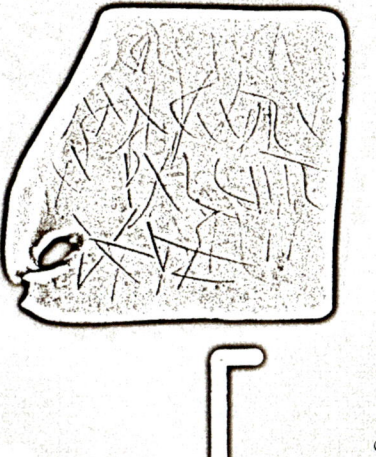
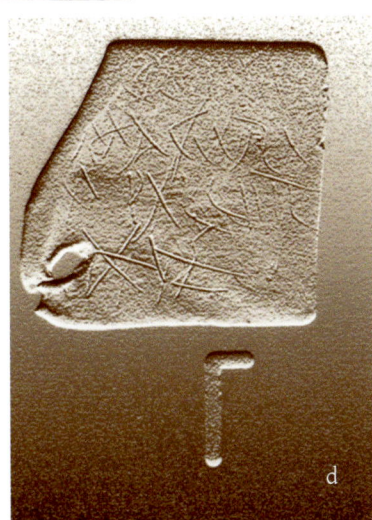
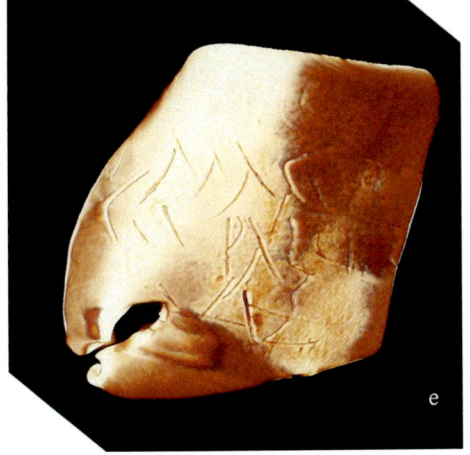

1st line	personal name	VERRUCILA	
2nd line	patronymic	MAGIRI	
3rd line	product	G(ausapae?) II PAL(lium) C(a) ER(uleum)	2 heavy coats? blue mantle
4th line	price	I £ X	1 denar and 1 dupondius

Reading (I. Radman-Livaja)

Figure 8: Inscribed lead label:
8a: non-restored price tag (Photo: M. Raab, ARDIG);
8b: Raw X-ray image (M. Schäfer, UFG Uni Wien);
8c: Reworked 6X-ray image
8d: Enhanced X-ray image
8e: Computer tomography (St. Handschuh, VETMED Wien);

cultivated crops such as cereals, (rye, barley, emmer wheat), pulses (lentil) and oilseeds (opium poppy, gold-of-pleasure) to exotic imports such as celery and muskmelon. Grapevine, of course – typical of a Roman site – is also present.

Identification of plant macroremains was complemented by pollen analysis carried out by Martina Weber (Department of Botany and Biodiversity Research, Division of Structural and Functional Botany, University of Vienna), the results of which may indicate the consumption of honey by the cesspit users.

The botanical remains from the cesspit's fill were also subjected to the AMS dating method, funded by the Austrian Archaeological Institute (OeAI) and carried out by Stephen Hoper und Paula Reimer (Queen's University Belfast). The result was that the samples had a 94% probability to date in a period between 50 and 220 AD.

The analysis of the parasites by Herbert Auer, Horst Aspöck, Ingrid Feuereis (Center for Pathophysiology, Infectiology and Immunology, Medical University of Vienna, Institute for Specific Prophylaxis and Tropical Medicine) could show that the eggs of *Ascaris lumbricoides* (roundworm), *Trichuris trichiura* (whipworm) and *Taenia* sp. (beef or pork tapeworm) were present, all in an extraordinary good state of preservation.

The coins and ceramics as well as the small finds will provide an insight into the social position of the users of the cesspit as well as helping to date the fill. Because the fill can be dated quite accurately, the ceramics will also be used to answer question on the range and development of ceramic forms as well as questions on production and trade. Similar questions will be answered with the help of the bone and metal finds and the beads will. The results will then be compared with the results from other findspots in Carnuntum and elsewhere.

The layer-for-layer excavation technique used during the excavation facilitated the reconstruction of the exact manner of the re-filling of the cesspit as well as the precise position of every find. The technical structure of the cesspit-latrine and its build, the inner encasing and the several filling layers were reconstructed three-dimensionally to allow further insights into this category of building. This is the first Roman cesspit-latrine to be published from the Austrian part of the province of Pannonia and the publication will be used as an opportunity to collect similar finds from excavations in the region.

The research project financed by the Austrian National Bank aims to present all features and finds and an interdisciplinary study of their regional and supra-regional context with the help of a team of researchers from different specialities.

The archaeological and scientific research will draw a picture of the everyday history – economy, food habits, health and hygiene – of this part of the province of Pannonia.

Acknowledgments

We thank the owner of Schloss Petronell for funding the excavation and for the possibility to scientifically examine and publish the finds

(Translated by Stefanie Hoss)

Bibliography

Petznek, B. 2014. Der Umgang mit Fäkalien in der römischen Antike. Ein kurzer Überblick anhand architektonischer, keramischer, literarischer und epigraphischer Quellen. In O. Wagener (ed.) *Aborte im Mittelalter und der Frühen Neuzeit. Bauforschung – Archäologie – Kulturgeschichte* (Freundeskreis Bleidenberg e.V. Studien zur internationalen Architektur- und Kunstgeschichte 117): 38–46.

Groh, S. and Petznek, B. 2012. Latrine in der Zivilstadt. ÖAI Wissenschaftlicher *Jahresbericht des Österreichischen Archäologischen Instituts 2011*: 68-69.

Petznek, B. 2012. Neueste Grabungen in Schloss Petronell 2010–2011. Vorläufige Grabungsergebnisse. *AÖ* 23/2: 29-34.

Radman-Livaja, I. 2014. *Tesserae Sisciensiae / Les plombs inscrits de Siscia* (Catalogues and Monographs 9).

Roman chamber pots

Beatrix Petznek

Chamber pots are large pottery vessels of (usually) oval shape, which for a long time had not been recognized correctly as chamber pots in the pottery of the Danube provinces of the Roman Empire. They were regarded as storage vessels, large cooking pots, wool baskets, misfired pottery, or even as bathing tubs for children. In contrast to this, they were described in French and Italian specialist publications as 'pitale' or 'pot de chambre'. (Frova 1977, Batigne-Vallet and Loridant 2000: 515-518, Pasqualini 2002, Bouet 2009.) In German-speaking pottery research, chamber pots were identified for the first time on the basis of the Carnuntum specimens. (Petznek and Radbauer 2008: 51-91, Petznek 2014: 38-46). Subsequently, chamber pots from Gallia Belgica, Germania and Pannonia were also recognized and presented (Bienert 2010, Hensen 2012, Chinelli 2011, Vámos 2014).

Chamber pots were used for defecation in private. That chamber pots were common in the rooms of Roman inns can be concluded from a graffito on the wall of a house in Pompeii. The host of an inn seemingly had refused to provide a chamber pot and found the following scratched onto the wall:

'miximus in lecto fateor peccavimus hospes/si dices quare nulla fuit matella.' –

'We have wet the bed, I admit. We have only sinned, o landlord, if you tell us for what reason there was no [urinal] pot.' (CIL IV 4957)

Chamber pots could be made from wood, clay or glass, but also of precious metals such as silver and gold. Luxurious vessels for urinating were confined to the rich upper classes. Ancient literature informs us about the different materials used for chamber pots, but only mentions the unusual – the normal, everyday is seldom noted. In his unique epigrams, Martial also talks about the materials of the chamber pots: *'Matella fictilis/ Dum poscor crepitu digitorum et verna moratur,/o quotiens paelex culcita facta mea est.'* – 'An earthenware chamber pot / As I am summoned by the snap of the fingers and the house slave dawdles,/ oh, how often the mattress has become my rival!' (Martial XIV 119, Transl. Loeb Classical Library, with amendments). *'Ventris onus misero, nec te pudet, excipis auro,/ Basse, bibis vitro: carius ergo cacas.'* – 'You receive your belly's load, Bassa, in gold—unlucky gold!—and are not ashamed of it; you drink out of glass. So it costs you more to shit.' (Martial I 37, Transl. Loeb Classical Library)

The Vita of Emperor Elagabal also expressly mentions the precious material of his chamber pots: *'Onus ventris auro excepit, in myrrinis et onychis minxit.'* – 'His belly's burden he received in gold, he pissed into [pots of] fluorite and onyx.' (HA vita Elagabalus 32, 3)

Chamber pots were also used for children. In Athens, a children's high chair with an integrated chamber pot was found. A similar model is depicted on an interior of an Attic bowl. (Pfuhl 1923, figs. 323, 326, Lynch and Papadopoulos, 2006: 1-32 with further comparisons, Neils and Oakley 2003: 148, cat 66; see Figures 1 and 2)

In addition to chairs with chamber pots for children, the ancient sources also mention such chairs for adults. These chairs had pots hung or placed under the bottom hole (see contribution Bienert in this volume).

Due to their relatively small size and weight, the chamber pots for defecating were easy to transport. Horaz writes of five servants who follow the praetor Tullius from the Tibur with his bottle basket and portable chamber pot (Sermonum I 107).

Literary sources also prove the emptying of chamber pots from the window onto the street in the dark of the night. The Digesten (9: 3) tell us that this was forbidden on pain of punishment. However, such a prohibition is proof that such a manner of disposal was quite common, as it seems to still be in some countries. (Thüry 2001: 25, fig. 29; http://investvine.com/singapores-problem-with-flying-garbage/ (retrieved 15-11-2016).

Juvenal claims that you can count yourself lucky if you are only hit by the contents of a chamber pot when walking the streets at night. (Satires III 268–277).

The finds of chamber pots are concentrated in civilian settlements, such as civilian towns, *canabae legionis* and *villae rusticae*. (see Appendix 1). In the military settlements, communal latrines were used (see Johnson 1987; Philipp 1997 45–56; Filgis 2005: 190–194, Ebeling, 124-127; Bishop 2012: 79, fig. 20). The extent to which these facilities were amended with chamber pots is unknown so far, but should not be underestimated according to Ebeling (2006: 124).

The contexts in which chamber pots were found are variable: latrines, bathhouses, drains and sewers, wells, port facilities, shipwrecks and potteries (see Appendix 2). These sites are directly related to the production,

Figure 1. Children's high chair with integrated chamber pot from Athens (after Lynch/Papadopoulos 2006).

Figure 2. Interior of an Attic bowl with depiction of children's high chair with integrated chamber pot (Sotades Painter, c. 460 BC, Musées Royaux d´Art et d´Histoire Brüssel, Inv.Nr. A890, Drawing: B. Petznek after Neils, Oakley 2003: 241, Cat. Nr. 42).

use and cleaning of the chamber pots or to the disposal of their contents.

In the Latin texts, several different names and forms are used for Roman chamber pots. The *matella* (Hilgers 1969, no. 230) is a jug-like vessel, a urine-bottle for the man. The use of simple pots or jugs for urinating is mentioned in literary sources and depicted in Greek vase painting (see Figure 3), but it is often impossible to identify these among the archaeological finds (Kolling 1993: 50, Pl. 77; Bouet 2009: 66, fig. 45 b).

> 'Cum peteret seram media iam nocte matellam/ Arguto madidus pollice Panaretus,/ Spoletina data est, sed quam siccaverat ipse,/ Nec fuerat soli tota lagona satis./ Ille fide summa testae sua vina remensus/reddidit oenophori pondera plena sui./ Miraris, quantum biberat, cepisse lagonam?/ Desine mirari, Rufe: merum biberat.' – ' When Panaret, maudlin, with snap of the thumb, /At midnight commanded the urinal to come;/ A spoletine came, which he himself had just drain'd:/ Nor had it sufficed that the flagon contain'd./ With utmost good faith redecanting his store,/ He crown'd the vast vessel as high as before./ Capacious, you wonder, the pot as the cask!/ This pure had imbibed; which accounts for the task.' (Martial VI 89 1, Transl. J. Elphinston)

Figure 3. Attic jug, Oinokles Painter around 470 BC in the Vatican depicting a satyr urinating into a jug (J. P. Getty Museum, Los Angeles, code 86.AE237, Drawing: B. Petznek after Neils, Oakley 2003: 148, Cat. Nr. 66).

Figure 4. Trulla from Rome, after Klauda (no year)

'*Digiti crepantis signa novit eunuchus/ Et delicatae sciscitator urinae/ Domini bibentis ebrium regit penem.*' – 'An eunuch knows the signal of a snapped finger,/ and, being the delicate expert of that urine,/ guides the penis of his boozy master.' (Martial III 82 15)

The *trulla* (Hilgers 1969, No. 364) is a not very deep vessel with a long handle and corresponds to today's bedpan. In Carnuntum, so far only relatively short handles with an attachment to the vessel's wall have been found (Petznek and Radbauer 2008: 79, pl. 8, 1-5), the complete form is therefore not known yet. However, a comparable chamber pot from Rome (Petznek and Radbauer 2008: 79, Pl. 8, 7 from Rome) or one from Pollentia (Vegas 1973: 38, fig 12.1) demonstrates the full form (see Figure 4).

The *scaphium* (Hilgers, 1969, no. 320) is a boat-shaped flat dish for the woman, which has not yet been archaeologically attested in Roman contexts. It corresponds to the modern Bourdalou. The name goes back to the then famous Jesuit priest Louis Bourdaloue, who gave his captive sermons at the court of King Louis XIV. In order for the ladies not have to leave the church during the sermon, they took along sauce boats for urinating.

'*Tolle, puer, calices tepidique toreumata Nili/ et milli secura pocula trade manu/ trita patrum labris et tonso pura ministro;/ anticus mensis restituatur honor. / Te potare decet gemma qui Mentora frangis/ in scaphium moechae, Sardanapalle, tuae.*' – Boy, take away these relief-decorated goblets from the warm Nile, and with nothing to fear hand me cups worn down by my ancestors' lips and a short-haired attendant to go with them; let old-world honour be restored to my table. It is appropriate for you, Sardanapallus, to drink from a bejewelled cup, you who break a [masterpiece by] Mentor to make a chamber pot for your mistress.' (Martial XI 11 5)

Characteristic of the large *lasanum* (see Figure 5) are the flat broad rim, oval mouth, large conical vessel body and the large and thus quite stable bottom. The vessel's height is adapted to a person's sitting posture during defecating. Only a small number of these chamber pots have handles. (Knossos, Aquitaine, Marsilia, Beneventum, Marcianopolis, Romula) Some have a widened seating area, in order to be able to better sit on them. (Vindobona, Drobeta, Romula, Knossos) No ceramic lids could be identified for the chamber pots yet. It is possible that wooden lids were used, or perhaps none at all. *Lasani* were already in use in ancient Greece as pictured in Greek vase paintings (Urinating *hetaira*: attic *kylix*, around 480 BC: Dierichs 1997: 119, fig. 133)

Literary sources also mention the *lasanum*: '*ab hoc ferculo Trimalchio ad lasanum surrexit. nos libertatem sine tyranno nacti coepimus invitare convivarum sermones.*' – ' After this dish Trimalchio got up [to go] to the *lasanum*. With the tyrant away we had our freedom, and we began to draw the conversation of our neighbours (Petronius Satyricon, Cena Trimalchionis II 41, 9)

These large, high chamber pots occur in Pannonia, Dacia, Moesia and Thrace, as well as in Northern Italy and Spain. There are four different types of vessels, which are found as early as the 1st century AD and are still produced in late pottery at the Carlino potteries (Magrini and Sbarra 2005: 136, pl. 54) as glazed chamber pots.

In the Gallic provinces, on the other hand, low bowls with straight or rounded walls were used from the 1st to the 3rd century AD. The rim is equally broad and flat, sometimes widened (Pompeii: Pasqualini 2002: 271, Figures 10, 11). Some of these vessels also have handles. Chamber pots as low as this require a squatting position to use.

Closer examination of the distribution of the two types of *lasani* (high and low, see Figure 6) reveals a correlation to the modern squat toilets (also known as Indian or Turkish toilets). The distribution of the low *lasani* in Gallia and Italia corresponds roughly to the distribution of the modern squat toilet culture of France and Italy.

Squatting toilets were already known in antiquity in the houses of Latium in Alba Fucens (Grassnick 1992 11, fig. 10, 11).

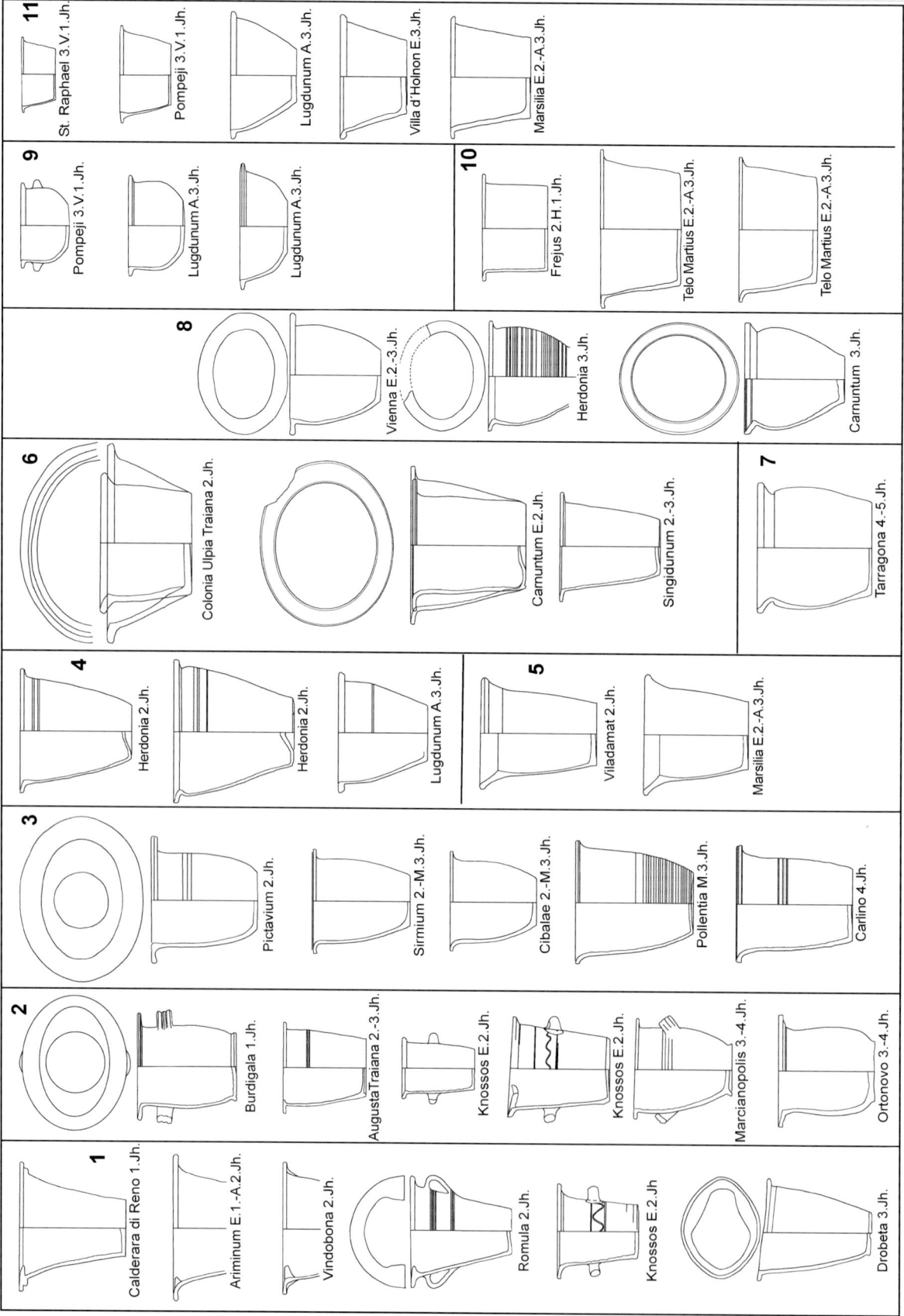

Figure 5: Different types of *lasani* (drawing. B. Petznek).

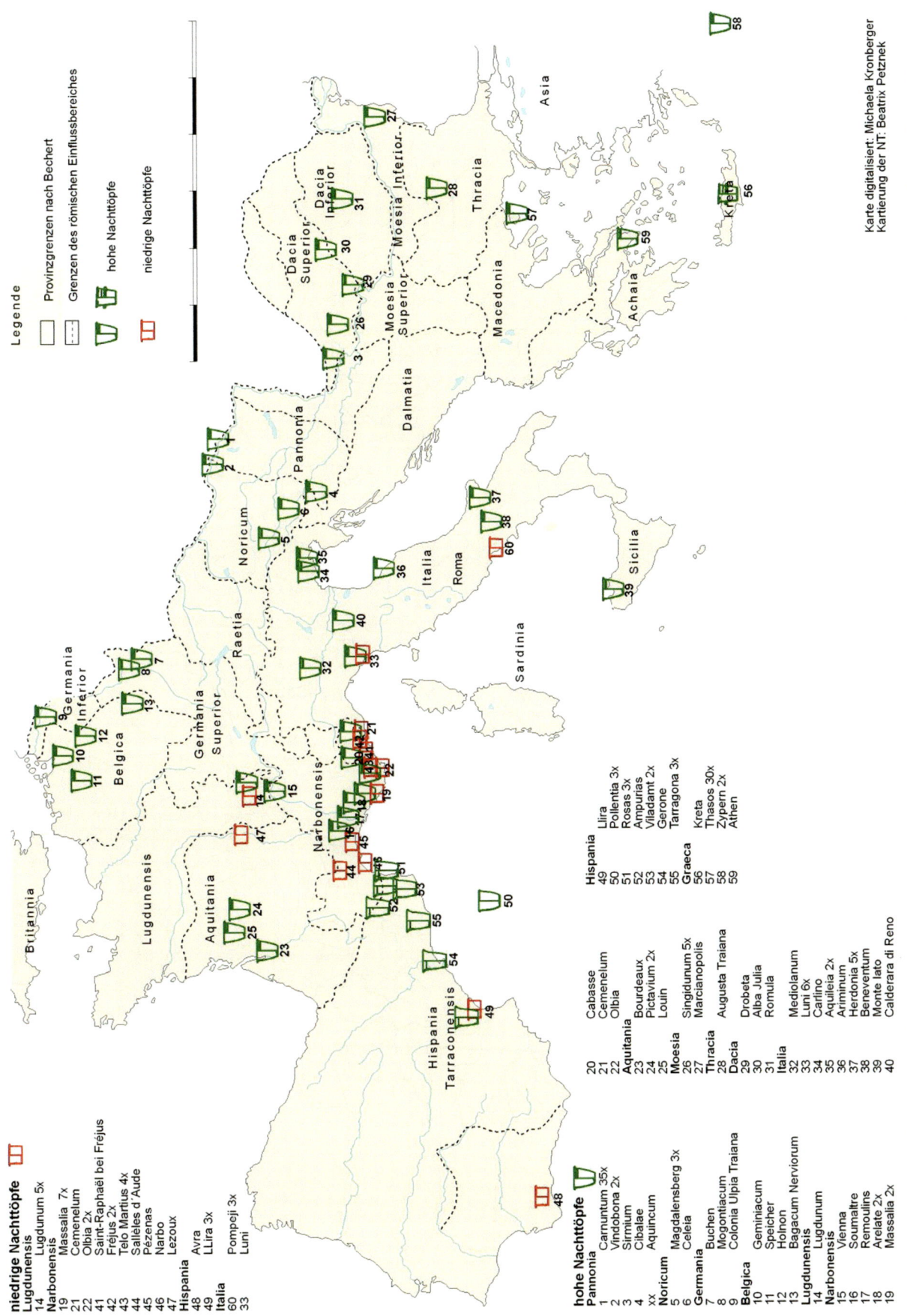

Figure 6: Distribution map of high and low chamber pot finds known to the author. (original map: Michaela Kronberger, distribution mapping: Beatrix Petznek).

Has this tradition continued since ancient times? Is this the reason why the French and Italians still use squatting toilets, whereas they are unknown in the Alpine region, for instance in the French-speaking western parts of Switzerland or the Italian-speaking population of Switzerland's Ticino? Who squatted for defecation and who sat?

The high *lasani* were made for sitting, while the on low ones, one squatted. In the large public latrines one sat, but on the low latrine holes in the street one could only squat – no indicators for a wooden seat were found.

Could this difference in preference have ethnic roots? Herodotus (On Egypt, Histories 2, 35) reported astonished of the different habits while peeing: 'with the Egyptians, the women pee standing up, the men seated.' (when using the *scaphium,* women could pee standing up – or rather, slightly squatting)

The distribution of the large, high chamber pots roughly corresponds to the settlement region area of the Celtic peoples – it seems that the Celts preferred sitting, while others, such as the Ligurians around Massalia in Gallia Narbonensis, preferred squatting. This difference in habit may have been connected to the type of dress these people wore. Men in long tunics have to squat when defecating, if they do not want to defile themselves. Men in trousers, however, have to sit down in order for their pants to not end up in the dirt. According to ancient writers such as Diodorus (5, 30, 1) the Celts '[...] wear patterned shirts of different colors and long trousers called *bracas*.' Strabon (4, 3, 3) affirms this, '[...] they dress in tight trousers.' Celtic men wore long trousers, while Greeks and Romans wore long tunics (Böhme-Schonberger 1997, Bieber 1977). This difference in clothing might be one of the reasons for the preference of certain population groups for the squatting or sitting while defecating.

Appendix 1: Chamber pots in civil contexts

Urban contexts
Pannonia: Carnuntum: Reinfuß 1961: 87 , 2/15; Reinfuß 1962: 60 Fig. 3/24,25; Petznek and Radbauer 2008, pl. 1-6; Vindobona: Petznek and Radbauer 2008, Pl. 7; Chinelli 2011: 67, fig. 66; Sirmium: Brukner 1981, pl. 126/174; Cibalae: Brukner 1981, pl. 126/173; **Noricum**: Magdalensberg: oral comm. Schindler-Kaudelka; **Moesia**: Singidunum: Bojović 1977, pl. 57/512; **Thrakia**: Augusta Trajana: Kalcev 1991: 261, fig. 11/10; **Italia**: Aquileia: Chinelli 1994: 321, pl. 50/ CCda 64,65; Mediolanum: Guglielmetti *et al*. 1991: 158, pl. 64/20; Monte Iato: Isler 1982: 219, pl. 79/3, fig. 7; Luna: Frova 1973: 770, pl. 213/1; Frova 1977: 187, pl. 125/3-5; **Hispania**: Lliria: Escrivà Torres 1994: 184; Bouet 2009, fig. 46/r´; Ampurias: Aquilué *et al.* 1984: 473; Bouet 2009, fig. 46/y´; Tarragona: Rüger 1968: 258, fig. 16/1-4; Pollentia: Vegas 1965: 120, fig. 5/1,2; Vegas 1973: 40, fig. 13/1; Rosas: Casas and Genoyer *et al.* 1990: 345; Bouet 2009, fig. 46/w´, x´; **Graeca**: Thasos: Gros 2014: 721, fig. 4/9; Zypern: oral comm. Schindler-Kaudelka; Athen: Robinson 1959, pl. 72; Bouet 2009, fig. 46/c´´; Mogontiacum: oral comm. A. Heising; Buchen-Rinschheim (Neckar-Odenwald-Kreis): oral comm. P. Mayer-Reppert; **Gallia Narbonensis**: Massalia: Bouet 2009, fig. 46/r; Moliner 1996: 241, fig. 7/1-3; Bouet 2009, fig. 46/k; **Belgica**: Geminiacum: Brulet *et al.* 2001: 337; Bouet 2009, fig. 46/l´;

Canabae legionis
Pannonia: Carnuntum: Grünewald 1983, pl. 43/1; Gassner 1990, pl. 3/18-21; **Moesia**: Singidunum: Nikolić-Đorđević 2000: 76, Typ II/24;

Villae rusticae
Gallia Narbonensis: Vienna: Godard 1995: 287, fig. 14/87; Remoulins: Fiches 1996: 353; Bouet 2009, fig. 46/u; Arelate: Bouet 2009, fig. 46/s,t; **Italia**: Herdonia: Annese 2000: 319, pl. 13/Typ 16/1-3; **Belgica**: Holnon: Batigne-Vallet and Loridant 2000: 517, fig. 2 below; Bouet 2009, fig. 46/m´; Bagacum Nerviorum: Batigne-Vallet and Loridant 2000: 517, fig. 2 above; Bouet 2009, fig. 46/k´; **Gallia Lugdunensis**: Lugdunum: Batigne-Vallet and Loridant 2000: 516 Fig. 1 below; Bouet 2009, fig. 46/g´,h´; **Gallia Aquitania**: Burdigala: Santrot and Santrot 1979, pl. 105, Typ 17, Form 443; Bouet 2009, fig. 46/y; **Hispania**: Viladamat: Casas and Genoyer *et al.* 1990: 345; Bouet 2009, fig. 46/z´,a´´; Girona: Casas and Genoyer et al 1990: 345; Bouet 2009, fig. 46/b´´; **Graeca**: Knossos/Kreta: Sackett *et al.* 1992, pl. 184/12, 195/103,104; Hayes 1983, fig. 10/117, 14/173,174, 15/175,176.

Appendix 2 : find contexts of chamber pots

Latrines
Italia: Herdonia: Annese et lal. 2000: 252-255, pl. 1/Typ 7/1-3; Emilia Romagna: Biondani 2005: 228, fig. 145/36;

Thermae
Gallia Narbonensis: Cemenelum: Grandieux 2004: 161, fig. 12/8, 13/1-5; Bouet 2009, fig. 46/a;

Drains
Pannonia: Petznek and Radbauer 2008;

Wells
Gallia Narbonensis: Olbia: Pasqualini 2002: 270, fig. 7; Bresciani und Excoffon 2004: 193, fig. 8/2; Bouet 2009, fig. 46/g;

Harbours
Gallia Narbonensis: Massalia: Pasqualini 2002: 271, fig. 9; Marty 2002: 216, fig. 12/60; Pasqualini 2002: 268,

fig. 3; **Italia:** Ariminum: Biondani 2005: 229, fig. 145/36,

Shipwrecks
Gallia Narbonensis: Massalia: Bouet 2009, fig. 46/n; Pasqualini und Pietropaolo 1998: 85; St. Raphael bei Fréjus: Pasqualini 2002: 268, fig. 2; Bouet 2009, fig. 46/e;

Potteries
Moesia: Marcianopolis: Mincev and Georgiev 1991: 235, fig. 17; **Dacia:** Drobeta: Stângă 1997: 627, fig. 5/7; **Noricum:** Vransko near Celeia: Vidrih-Perko 1997: 165, fig. 2/9; **Italia:** Beneventum: Pasqualini 2002: 272, fig. 12; Bouet 2009: 68, fig. 46/p´; Carlino: Magrini and Sbarra 2005: 136, pl. 54; **Pannonia:** Aquincum: Vámos 2014, Katnr. 623-631, pl. 57-58; Láng et al. 2014: 71; **Gallia Narbonensis:** Aspiran: Mauné 2001: 179; Bouet 2009: 68, fig. 46/v; **Belgica:** Speicher: Bienert 2010: 50, fig. 1-4 (see Bienert, this volume); **Dacia:** Romula: Popilian 1997, pl. 46/5;

Pottery shop
Gallia Aquitania: Pictavium: Wittmann and Jouquand 2003: 636, fig. 17/63; Bouet 2009, fig. 46/a´,b´; Louin: Robin and Chambon 2002: 367; Bouet 2009: 68, fig. 46/c´).

Necropolis
Gallia Narbonensis: Cabasse La Calade: Bèrard 1961, Grave 22/127, 137, pl. 23; Bouet 2009: 68, fig. 46/b; Massalia Sainte-Barbe: Moliner 1996: 241, fig. 7/6; Pasqualini 2002, fig. 8; Bouet 2009: 68, fig. 46/m; **Germania:** Colonia Ulpia Traiana: Hensen 2012: 10.

Bibliography

Sources

Historiae Augustae. Translated Hohl, E. 1976. *Historia Augusta Römische Herrschergestalten*. Bd. 1 von Hadrianus bis Alexander Severus, München.

Secondary literature

Annese C., De Felice, G. and Turchiano, M. 2000. Ceramiche della prima e media età imperiale dai riempimenti delle latrine della domus A. In Volpe G., *Ordona 10. Ricerche archeologiche a Herdonia 1993-1998*: 251-265. Bari.

Aquilué Abadías, X., Mar Medina, R., Nolla Brufau, J. M., Ruiz de Arbulo Bayona, J. and Sanmartí Grego, E. 1984. *El fòrum romà d'Empúres, Excavacions de l'any 1982. Un aproximació arqueològica al procés històric de la romanizació al nord-est de la península ibèrica*. (Monografies emporitanes 6). Barcelona.

Batigne-Vallet C. and Loridant, F., 2000. Notes sur les seaux de Bavay et les récipients ovalisés de Lyon, Société Française d'Étude le Céramique Antique en Gaule (SFECAG) Actes Cougrès de Libourne: 515-518.

Bendi, C. 2005. Ceramica comune depurata. In Ortalli J., Poli P., Trocchi T., *Antiche genti della pianura. Tra Reno e Lavino. Ricerche archeologiche a Calderara di Reno. Quaderni di Archeologia dell'Emilia Romagna* 4: 54-63.

Bèrard, G. 1961. La nécropole gallo-romaine de La Calade à Cabasse (Var). *Gallia* 19/1: 105-158.

Bieber, M. 1977. *Griechische Kleidung*. Berlin, New York.

Bienert, B. 2010. Vasa obscena – antike Sanitärkeramik. Zu einer römischen Klosettschüssel aus Speicher, Eifelkreis Bitburg Prüm. *Funde und Ausgrabungen im Bezirk Trier* 42: 49-63.

Biondani, F. 2005. Ceramica comune decorata acroma. In Mazzeo Saracino L., *Il complesso edilizio di età romana nell'area dell'ex vescovado a Rimini*: 222-234. Sesto Florentino.

Bishop, M. C. 2012. *Handbook to Roman Legionary Fortresses*. Barnsley.

Böhme-Schönberger, A. 1997. *Kleidung und Schmuck in Rom und den Provinzen* (Schriften des Limesmuseum Aalen 50). Stuttgart.

Bojović, D. 1977. *Rimska keramika Singidunuma: die römische Keramik von Singidunum* (Serija-Zbirke i legati muzeja grada Beograda. Katalozi 8). Beograd.

Bouet, A. 2009. *Les latrines dans les provinces gauloises, germaniques et alpines* (Gallia Supplement 59).

Brukner, O. 1981. *Rimska keramika u jugoslavenskom delu provincije Donje Pannonije*. (Roman ceramic ware in the Yugoslav part of the Province of Lower Pannonia). (Dissertationes et Monographie XXIV) Belgrade.

Brulet, R., Dewert, J. P. and Vilvorder, F. 2001. *Liberchies IV, Vicus gallo-romain, Travail de Rivière, Fouilles du musée de Nivelles 1986/87 et 1991/97)*, Louvain-la-Neuve.

Casas, J., Castanyer, G. P., Nolla, J. M. and Tremoleda, J. 1990. *Ceràmiques comunes i de producció local a les comarques orientals de Girona. I. Materials augustals i alto-imperials*. (Centre Investigacions Arqueològiques de Girona. Sèrie Monogràfica 12). Girona.

Chinelli R. 1994. Ceramica commune. In M. Verzar Bass (ed.), *Scavi ad Aquileia I. L'area ad Est del foro. Rapporto degli scavi 1989-1991*: 321-323. Roma.

Dierichs, A.,1997. *Erotik in der Kunst Griechenlands*, Antike Welt Sonderheft. Mainz.

Ebeling, Chr. 2006. Les latrines. In M. Reddé, R. Brulet, R., Fellmann, J. K. Haalebos, and S. von Schnurbein, *Les fortifications militaries, L' architecture de la Gaule romaine* (Documents d'archéologie française 100): 124-127. Bordeaux.

Escrivà Torres, V., 1994. *Cerámica común romana del Municipium Liria Edetanorum. Nuevas aportaciones al estudio de la cerámica de época alto-imperial en la Hispania Tarraconensis, Cerâmica comuna romana d'època alto-imperial a la península ibèrica*. Estat de la qüestió (Monografies Emporitanes VIII, Museu d'Arqueologia de Catalunya-Empúries): 167–186. Ampurias.

Fiches, J. L. 1996. Céramiques culinaires et vaisselle commune de table dans la région de Nîmes. In M. Bats (ed.), *Les céramiques communes de Campanie et de Narbonnaise (fin Ier s. av – IIe s. ap. J.-C.): la vaisselle de cuisine et de table, Naples mai 1994*: 351–359. Naples.

Filgis, M. N. 2005. Wasser und Abwasser. Infrastruktur für Soldaten und Bürger. In *Imperium Romanum. Roms Provinzen am Neckar, Rhein und Donau*: 190–194. Esslingen.

Frova, A. (ed.) 1977. *Scavi di Luni II. Relazione preliminare delle campagne di scavo 1972-1973-1974*. Rom.

Gassner, V. 1990. Gelbtonige Keramik aus datierten Fundkomplexen in Carnuntum. *Carnuntum Jahrbuch* 1989: 133–161

Godard, C. 1995. Quatre niveaux d'abandon de la ville de Vienne (Isère): Eléments pour la chronologie des céramiques de la fin du IIe siècle et du IIIe siècle après J.-C. *Société Française d'Étude le Céramique Antique en Gaule (SFECAG) Actes Congrès de Rouen*: 285–321.

Grandieux, A. 2004. La céramique commune de l'espace sud des Thermes de l'Est de *Cemenelum* à Nice/Cimiez (Alpes-Maritimes). Un contexte du Haut-Empire et de l'Antiquité tardive. *Societé Française d'Étude de la Céramique Antique en Gaule (SFECAG) Actes du Congrès de Vallauris*: 151–165.

Gros, J. S. 2014. Une forme particulière des abords de l'agora des Thasos: La bassine ovale à Marli. In N. Poulou-Papadimitriou, E. Nodarou and V. Kilikoglou (eds) *Late Roman coarse wares, cooking wares and amphorae in the Mediterranean. A market without frontiers (LRCW 4) (BAR International Series 1216)*: 715–722. Oxford.

Grünewald, M. 1983. *Die Funde aus dem Schutthügel des Legionslagers von Carnuntum (Die Baugrube Pingitzer) (Der römische Limes in Österreich 32)*. Wien.

Guglielmetti, A., Lecca Bishop, L. and Ragazzi, L. 1991. Ceramica comune. In D. Caporusso (ed.), *Scavi MM3. Ricerche di archeologia urbana a milano durante la costruzione della linea 3 della Metropolitana 1982-1990*, Bd. 3.1: 133–258. Milano.

Hayes, J. W. 1983. The villa Dionysos excavations Knossos: the Pottery. *The Annual of the British School at Athens* 78: 97–169.

Hensen, A. 2012. Stille Örtchen im Römischen Reich. *Archäologie in Deutschland* 1: 8–13.

Hilgers, W. 1969. Lateinische Gefäßnamen. Bezeichnungen, Funktion und Form römischer Gefäße nach den antiken Schriftquellen (Beihefte Bonner Jahrbücher 31). Düsseldorf.

Isler, H. P. 1982. Eine Fundgruppe des 5. Jahrhunderts n. Chr. aus der Siedlung auf dem Monte Iato. *Mitteilungen des Deutschen Archäologischen Instituts Römische Abteilung* 89: 213–225.

Johnson, A. 1987. *Römische Kastelle des 1. und 2. Jahrhunderts n. Chr. in Britannien und in den germanischen Provinzen des Römerreiches (Kulturgeschichte der antiken Welt 37)*. Mainz am Rhein.

Kalcev, K. 1991. Zur Herstellung der antiken Keramik in Augusta Trajana/Stara Zagora. *Rei Cretariae Romanae Fautorum Acta* 29/30: 245–273.

Klauda, M. (no date) *Erstes Nachttopf-Museum der Welt, Zentrum für Außergewöhnliche Museen (ZAM Katalog 14)*. München.

Kolling, A. 1993. *Die Römerstadt in Homburg-Schwarzenacker*. Homburg.

Láng, O., Nagy, A. and Vámos, P. 2014. *The Aquincum macellum. Researches in the area of the macellum in the Aquincum Civil Town 1882-1965 (Aquincum Nostrum I.3)*. Budapest.

Lynch, K. M. and Papadopoulos, J. K. 2006. Sella cacatoria. A study of the potty in archaic and classical Athens. *Hesperia* 75: 1–32.

Magrini, Ch. and Sbarra, F. 2005. *Le ceramiche invetriate di Carlino. Nuovo contributo allo studio di una produzione tardoantica*. Florence.

Marty, F. 2002. Aperçu sur les céramiques à pâte claire du Golfe de Fos/Marseille. In L. Rivet and M. Sciallano (eds), *Vivre, produire et échanger. Reflets méditerranéens: mélanges offerts à Bernard Liou (Archéologie et histoire romaine 8)*: 201–220. Montagnac.

Mauné, S. 2001. Les ateliers de potiers d'Aspiran. Nouvelles données et perspectives. In F. Laubenheimer, (ed.), *Actes du colloque international de Sallèles d'Aude Le monde des potiers gallo-romains, 27 et 28 septembre 1996*: 159–194. Paris.

Mincev, A. and Georgiev, P. 1991. Marcianopolis. Ein neues Zentrum der Keramikproduktion im 2.-6. Jahrhundert. *Rei Cretariae Romanae Fautorum Acta* 29/30: 223–244.

Moliner, M. 1996. Les Céramiques communes à Marseille d'après les fouilles récentes. In M. Bats (ed.), *Les Céramiques communes de Campanie et de Narbonnaise (Ier s. av. J.-C. – IIe s. ap. J.-C.): la vaisselle de cuisine et de table. Journées d'étude. (Collection du Centre Jean Bérard 14)*: 237–255. Naples.

Müller, M., Chinelli, R., Czeika, S., Mader, I., Jäger-Wersonig, S., Sakl-Oberthaler, S., Eisenmenger, U., Litschauer, C., Öllerer, C. and Eleftheriadou, E. 2011. *Entlang des Rennwegs. Die römische Zivilsiedlung von Vindobona (Wien Archäologisch 8)*. Wien.

Neils, J. and Oakley, J. H. (eds), 2003. *Coming of Age in ancient Greece. Images of Childhood from Classical Past*. New Haven.

Nikolić-Đorđević, S. 2000. Antička keramika Singidunuma – oblici posuda, *Singidunum* 2: 11–206.

Pasqualini, M. 2002. Le pot de chambre, une forme particulière du vaisselier céramique dans la maison romaine entre les Ier et IIIe siècles de notre ère. In L. Rivet and M. Sciallano, M. (eds), *Vivre, produire et échanger. Reflets méditerranéens. Mélanges offerts à*

Bernard Liou (Archéologie et histoire romaine 8): 267–274. Montagnac.

Pasqualini, M. and Pietropaolo, L., 1998. Le secteur de l'épave, Les céramiques communes d'origine indéterminée. In M. Bonifay, M.-B. Carre and Y. Rigoir (eds), *Fouilles à Marseille: contextes et mobiliers (Ier-VIIe siècles)* (Etudes Massaliètes 5): 80-93.

Petznek, B. and Radbauer, S. 2008. Römische Nachttöpfe aus der Zivilstadt von Carnuntum. *Carnuntum Jahrbuch* 2008: 51–91.

Pfuhl, E. 1923. *Malerei und Zeichnung der Griechen.* München.

Philipp, M. 1997. Die Grabungen in der südöstlichen Ecke des Kastells. Ein Vorbericht. In M. Kandler (ed.), *Das Auxiliarkastell Carnuntum 2. Forschungen seit 1989* (Sonderschriften des Österreichischen Archäologischen Instituts 30): 45–56. Wien.

Popilian, G. 1997. Les centres de production céramique d'Olténie. In D. Benea (ed.), *Études sur la céramique romaine et daco-romaine de la Dacie et de la Mésie Inférieure* (Bibliotheca Historica et Archaeologica Universitatis Timisiensis 1): 7–20. Timisoara.

Reinfuß, G. 1961. Die Keramik der Jahre 1958/59, *Carnuntum Jahrbuch* 5: 74–99

Reinfuß, G. 1962. Keramik der Jahre 1953/54. *Carnuntum Jahrbuch* 6: 54–95.

Robin, K. and Chambon, M. P. 2002. La Martinière (Deux-Sèvres): un atelier de potiers du Bas Empire. *Aquitania* XVIII: 343–371

Robinson, H. R. 1959. *The Athenian Agora V, Pottery of the Roman Period: Chronology.* Princeton.

Rüger, Chr. 1968. Römische Keramik aus dem Kreuzgang der Kathedrale von Tarragona. *Madrider Mitteilungen* 9: 237–258.

Sackett, L. H. (ed.) 1992. *Knossos from Greek City to Roman Colony. Excavation at the Unexplored Mansion II* (British School of Archaeology at Athens Supplement 21). Athen.

Santrot, M. H. and Santrot, J. 1979. *Céramiques communes gallo-romaines d'Aquitaine.* Paris.

Stângă, I. 1997. Un centre céramique dans le territoire rural de Drobeta. *Acta Musei Napocensis* 34: 621–634.

Vámos, P. 2014. Majdnem terra sigillata, Adatok az aquincumi canabae katonai fazekasműhelyének legkorábbi periódusához. (Fast Terra Sigillata. Angaben über die frühesten Periode die militärische Töpferei in den Canabae von Aquincum). In P. Balázs (ed.), *Fiatal Római Koros Kutatók* III. konferenciakötete: 143-160. Szombathely.

Vegas, M. 1965. Spätkaiserzeitliche Keramik aus Pollentia (Pollèntia Mallorca). *Bonner Jahrbuch* 165: 108–140.

Vegas, M. 1973. *Cerámica común romana del mediterráneo occidental.* Barcelona.

Vidrih-Perko, V. 1997. The Roman Tile Factory At Vransko Near Celeia (Noricum) Part Two: Ceramic Finds, *Rei Cretariae Romanae Fautorum Acta* 35: 165–172.

Volpe, G. 1998. Archeologia subacquea e commerci in età tardoantica. In G. Volpe (ed.) *Archeologia subacquea. Come opera l'archeologo sott'acqua. Storie dalle acque:* 561-626. Firenze.

Wittmann, A. and Jouquand, A. M. 2003. La boutique d'un marchand de vases dans la seconde moitié du IIIe siècle après J.-C. à Poitiers (Vienne), *Société Française d'Étude le Céramique Antique en Gaule (SFECAG) Actes du congrès de Saint-Romain-en-Gal:* 621-639.

A Roman 'Toilet bowl' from Speicher
(Eifelkreis Bitburg-Prüm, Rhineland-Palatinate, Germany)

Bernd Bienert

The Roman potteries southeast of Bitburg in the forest between Speicher and Herforst own their existence to the large tertiary clay deposits located there. From the second century AD onwards, the potters of Speicher used them to produce enormous amounts of coarse ware. The different forms include jugs, jars, basins, large and small bowls, platters, pots, beakers and lids. Using untempered white burning clay allowed the imitation of metal vessels like jugs, plates and trays. Jars, bowls, basins, plates and beakers covered in a red burning slip or engobe and occasionally painted in white take fine Samian ware (Terra Sigillata) as a model. In the 3rd century AD, the potters turned to marbling the outside of drinking vessels, jugs and jars with red burning slip. In the 4th century, drinking vessels, jars, jugs and vessels with several spouts were decorated with lamellae-like patterns running top to bottom, in the style of red flamed ware (Loeschcke 1922: 10-13, fig. 9-11). The range of special forms in this ware include pots used in oven-building (Wölbtöpfe), face urns, bowls for making cheese, money boxes, incense burners, candle holders, oil lamps and crucibles. Another form can now be added to this list – the toilet bowl.

The vessel described here belonged to the collection of Jacob Plein-Wagner and has been described earlier by the author (Bienert 2001: 41, No. 41). It will be displayed in the local museum at Speicher in the future. The vessel is high, with a simple round and flat bottom (width bottom 12 cm, height 21,3 cm, width opening 30 cm, breadth opening 26,3 cm) (Figure 1). The vessel widens conically from the bottom to the mouth, the outside is completely smooth and without any structure, e.g. grooves. Around the oval mouth runs a wide horizontal rim, which is slightly convex on top and flat on the bottom (Figure 2).

The form of the vessel's body and the oval form of the mouth are in no doubt even though the vessel has been assembled from sherds. Missing pieces in the vessel wall and the rim and mouth (the latter being about a sixth) have been supplemented with plaster. To stabilize the whole, the inside was completely covered in plaster. While this support ensured the position of the sherds, it did lead to some inexactness.

The clay used was cream-colored, and quartz sand was used as temper. The latter is quite homogenous in grain size and composition. The yellowish to orange outer layer is typical for Speicher Ware, which was burned moderately hard at 900° C. The vessel's body has been quickly built up from the wheel without any flourishes. Tool-traces manifesting

Figure 1. Speicher, Eifelkreis Bitburg-Prüm, Germany. Coarse ware toilet bowl from above (photo: Alain Anfossy, Dudeldorf, and Bernd Bienert, Trier).

Figure 2. Speicher, Eifelkreis Bitburg-Prüm, Germany. Coarse ware toilet bowl, detail mouth (photo: Alain Anfossy, Dudeldorf, and Bernd Bienert, Trier).

Figure 3. Speicher, Eifelkreis Bitburg-Prüm, Germany. Coarse ware toilet bowl, detail bottom (photo: Alain Anfossy, Dudeldorf, and Bernd Bienert, Trier).

themselves as concentric loops on the underside were caused by the use of a twisted cutting wire (Figure 3). The traces prove that the bottom had not been reworked after being cut from the wheel. The mouth had originally been formed circular on the wheel, but was then made oval by slight pressure. As only one fingerprint has been found on the outside, the vessel's body must have been smoothed with a sponge or cloth afterwards. Because of the form of restoration chosen, the original thickness of the wall cannot be judged.

The precise findspot of the Speicher toilet bowl is unknown, however, the possibilities can be narrowed down. The unconventional manner of restoration proves that the vessel belonged to the collection of Jacob Plein-Wagner, whose collection was formed between the years 1876/77 and 1903. It is known that he caused excavations on three places near Speicher at the locations called '*Herst*', '*Pützchen*' and '*Herforster Wäldchen*' (findspots number 13, 20 and 21 resp., Bienert 2001: 17). The Speicher toilet bowl must have come from one of these three findspots.

The dating of the vessel is based on the following considerations: The remaining inventory of the collection Jacob Plein-Wagner points to the fact that during the 19th century, only the upper layers of the refuse dumps had been dug. The misfired pottery from these layers had been manufactured in the latest production phase of the Speicher potteries. Consequently, it can be assumed that the toilet bowl may be dated into the 4th or the early 5th century AD.

High vessels with an oval mouth are an absolute rarity in the extensive repertoire of forms from the Speicher potteries. The vessel from the collections of Jacob Plein-Wagner presented here is the only known example. It must remain uncertain if the vessel was produced for sale or if it is a singular piece for use at the potter's own home. During its first publication, the vessel was interpreted as a 'wool-basket' (*calathus*) and because of its rarity was thought to have been a special production for commercial purposes. As the function of these vessels is clear now, this interpretation has been rendered obsolete.

In his publication on latrines in the Gallic, German and Alpine provinces of the Roman Empire, Alain Bouet presents similar vessels found in Bavay (Dép. Nord), Cabasse (Dép. Var), Holnon (Dép. Aisne), Hyères-les-Palmiers (Dép. Var), Llíria (Catalonia, Spain), Louin (Dép. Deux-Sèvres), Lyon (Dép. Rhône), Marseille (Dép. Bouches du Rhône), Nizza (Dép. Alpes-Maritimes), Remoulins (Dép. Gard), Soumaltre (Dép. Hérault) and Toulon (Dép. Var) (Figure 4). In addition, other types exist, which have a straight, curved or s-curved shape. Some have been designed as pots, others as bowls. They may have handles formed either by lobes on the rim

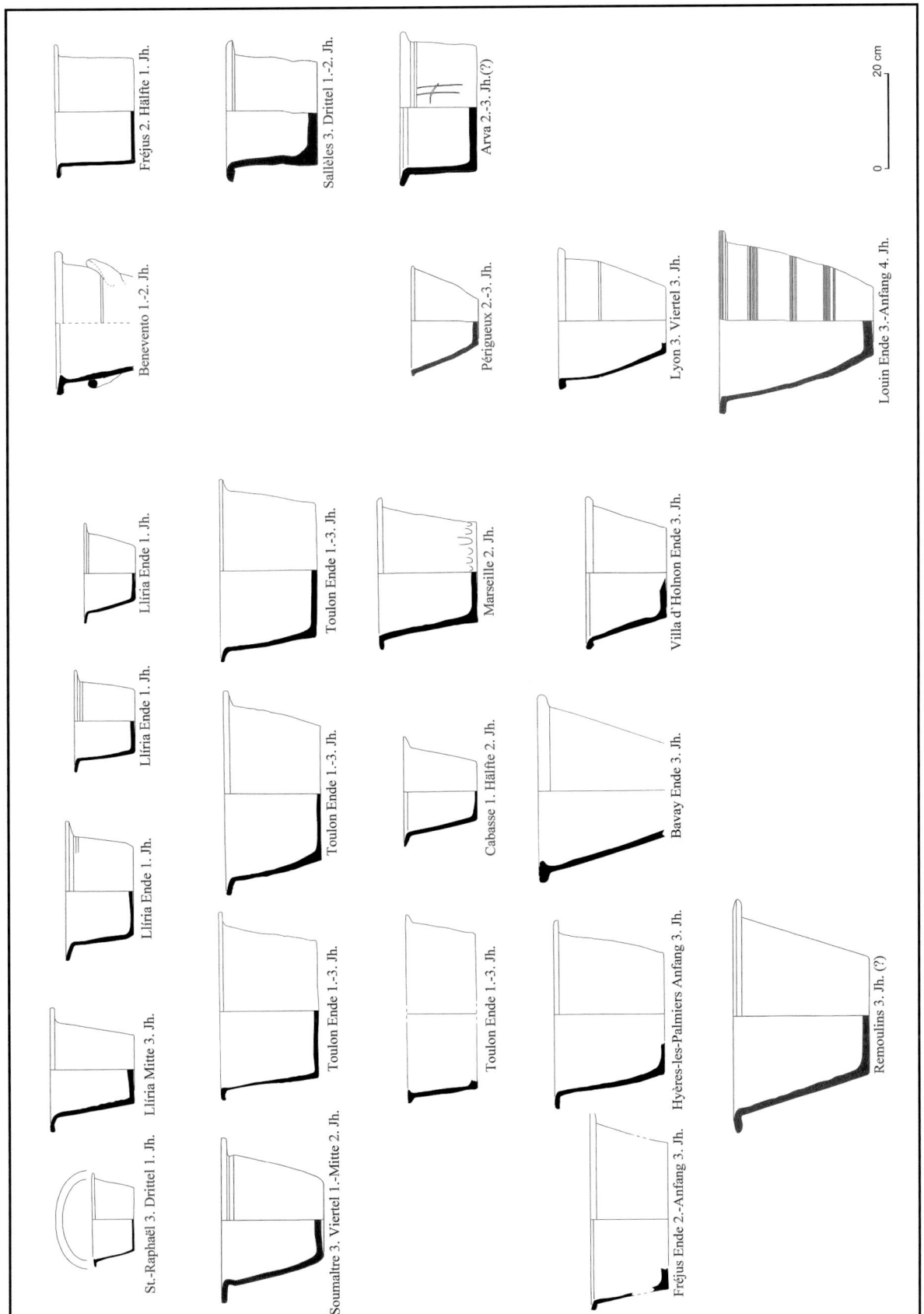

Figure 4. Distribution and dating of conical vessels with oval mouth (after Bouet 2009: 69-72 fig. 46a-c).

or by handles on the shoulder or they may be without handles. The rims around the mouths are either horizontal, tilt upwards or are simply folded out. The most striking feature common to all types is the oval mouth. Lids belonging to these vessels have hardly ever been found. Interpreted as chamber pots, their usage covers a period from the early 1st to the early 6th century AD (Bouet 2009, p. 68-75, fig. 46a-c).

The form of the conical vessel shows similarities to another Roman vessel form: oval bronze bowls with a lid, horizontal rim, movable handles and pelta-shaped feet. Due to the formal similarities, it can be assumed that these were chamber pots as well (Bouet 2009: 73, fig. 47a-r).

As for their ceramic counterparts, new findings from the civil town of Carnuntum (Archäologischer Park Carnuntum, Bad Deutsch-Altenburg, Lower Austria) unambiguously demonstrate the use of these vessels. During excavations on the Western Street in the years 2002-2003, around 30 vessels were found in the sewage system. Among them were several with a flat bottom, conical vessel shape, oval mouth and wide rim. On their inside, deposits of 0,5 to 1,5 mm thickness were discovered, which were made up of alternating irregular layers coloured whitish-yellow and dark brown. Mineralogical and petrographical analyses demonstrated that these layers consisted of urine scale (R. Sauer in: Petznek/Radbauer 2008: 71). Consequently, the vessels found in Carnuntum can be identified with certainty as sanitary vessels (*vasa obscena*; *vasa turpia*).

The height of the conical vessels varies between 12 and 34 cm, the width of the oval mouth between 20 and 48 cm. Their characteristics - firm stand, high wall, oval mouth and wide rim - point to their use in a sitting or squatting position. Different measurements speak of standardized vessels for the use of children and grown-ups or one person versus several.

These large conical vessels have parallels in Germany, France, Italy and Spain, from the Balkans and Crete (Bouet 2009: 68-73; Petznek/Radbauer 2008: 51-91). As settlement ceramics, they were found in civil towns, settlements around legionary fortresses (*canabae legionis*) and country estates (*villae*). They have been discovered in filled-up latrines, sewers, wells and cisterns, as well as bath buildings, ports, wrecked ships and commercial businesses. An association with potters' workshops was evident several times (Bouet 2009: 68-73; Petznek/Radbauer 2008: 60). Understandably, the vessels were seldom used as grave gifts, but examples for this use come from Cabasse, La Calade and Marseille, Sainte-Barbe (Bouet 2009: 68, fig. 46b and 46m).

According to the written sources, the *lasanum* - colloquially *lasanus*, Ancient Greek *lasanon* – was intended for the reception of human excrement (Helenius Acron, Commentarii in Q. Horatium Flaccum. In sermonem I 6, 109-111). The word *lasanus* is attested as a derogatory term in the Anthologia Latina (205, 13). The numbers of written sources on these vessels are small and include only rare clues enabling a reconstruction of the form. Martial (Epigrammata I 37, 1-2) and Aelius Lampridius (Heliogabal 32, 2) mention examples made from gold. In both cases though, the fact is used as a hyperbole and may thus not record ancient use. In his book on the interpretation of dreams, Artemidoros reports the case of someone who has dreamed of defecating into a *choinix*, a vessel for dry measures (Oneirokritika V 24). This vessel may have had a similar shape to another measuring vessel for grain, the *modius*, on whose shape we are much better informed, as it was among other things an attribute of the god Sarapis (Hilgers 1969: 67-68, 224-225, No. 245). A comparison of the forms allows the hypothesis that the *choinix* was an allusion to the cylindrical and conical shape of the toilet bowls.

In order to use *lasana*, one discreetly withdrew into a room comparable to what we call a toilet today. According to Petronius, Trimalchio did not prevent anyone to relieve himself within the dining room (*triclinium*). 'But even if the matter is serious, everything is ready outside: water, *lasani* and *minutalia*' (Cena Trimalchionis 47, 5). The latter will have included the sponges on sticks, which – together with pieces of cloth – were used to clean oneself. It seems likely that *lasana* were placed in toilets (*latrinae*) and sleeping rooms (*cubicula*). The dream of amply defecating while sitting on a *lasanum* is judged a positive omen by Artemidoros (Oneirokritika II 26). This is proof that these toilet vessels were not used in a squatting, but a sitting position. The reinforcement that some types carry on the inside of the mouth strengthens the case of their being used in a sitting position (Bendi 2000: 52, pl. 1,9; Hayes 1983: 132, fig. 14,173; Petznek/Radbauer 2008: 56, pl. 7,1-2). Perforated sections of the rim as well as handles in the shape of lobes indicate that the relevant vessels were hung into (wooden) toilet chairs or stools (Batigne-Vallet/Loridant 2000: 518, fig. 3; F. Vilvorder in: Brulet et. al. 2001: 335-336, fig. 137,1-2, 6-7; Herbin et al. 2004: 45, fig. 12,88). Horatius and several Greek sources testify to the use of transportable *lasana* (Sermones I 6, 109), which were carried by slaves – *lasanophoroi* – to their masters and must have been recognizable from far off (Plutarch, Moralia 182 c and 360 d). As a matter of course, occupations like carrying the *lasana*, holding the urine bottle (*matellae*) or cleaning the latrines were seen as highly demeaning.

The *sella pertusa* is mentioned in connection with urine as medication and described as a chair with a perforated sitting area (Cato, De agri cultura 157, 11). Just a small amount of stone versions of these chairs have been discovered up to now (Grassnick 1992: 8-10,

Figure 5. Munich, Germany, Zentrum für Außergewöhnliche Museen: Childern's chamber pot integrated into a portable rattan box. Southern Germany, about 1830 (after Klauda 1986: 12 fig. below on the left side).

fig. 6-9; Richter 1966: 99-100, fig. 503-504; Smith 1904: 403, No. 2517, fig. 65). Their keyhole-shaped hole reaching to the bottom shows that they were installed above a channel. This makes it likely that they were used in latrines, similar to the one, which was found in the *domus Augustana* on the Palatine hill in Rome in the 18th century (Bouet 2009: 64, fig. 44c and 75).

Ceramic bowls with oval mouths and conical shape are most likely the *lasana* mentioned in the sources. Their usage was twofold: either freestanding with the person squatting over the vessel or as an inset for (wooden) toilets chairs standing in a room, which were used in a sitting position.

Vasa obscena are often seen as chamber pots, used as a makeshift measure at night. But their usage is much wider than that of the chamber pot. The *matella* can be identified as the urine bottle for men, the *scaphium* as *vas urinae* for women and the *lasanum* as toilet bowl. The latter consists of vessels with a round flat bottom, cylindrical-conical shape and oval mouth, whose form and measurements are comparable to modern toilets. In analogy to these, *lasana* can be interpreted as insets for stationary (Petronius, Cena Trimalchionis 41, 9 and 47, 5) or movable seats (Horaz, Sermones I 6, 109; Plutarch, Moralia 182 c and 360 d) (Figure 5). The integration into a chair or seat would also explain the lack of a handle and the rarity of finds with a lid.

The effectiveness of the form is proven by its long and almost unchanged use over several centuries. The small number of finds seems to contradict this fact at first. But this can be countered by pointing out that the amount of finds of this type are mainly the result of the last two decades of research. Another difficulty may be to recognize the sherds of types with oval mouths among the settlement ceramics and not ascribe them to misfired pieces, but to *lasana*.

Aelius Lampridius reports that emperor Heliogabal (217-222 n. Chr.) used golden vessels for defecating and urinated into vessels of fluorite and onyx (Scriptores historiae Augustae, Heliogabal 32, 2). The toilet bowl of Speicher cannot keep up with such luxurious products. However, its discovery is a small sensation in itself, as it demonstrates the use of *lasana* in these parts. And – as the connection to the pottery shop is quite clear – this proves for the first time that the potters of Speicher produced these toilet bowls.

(Translated by Stefanie Hoss)

Bibliography

Batigne-Vallet, C. and Loridant, F. 2000. Note sur les seaux de Bavay et les récipients ovalisés de Lyon. In *Actes du congrès de Libourne 1.-4. juin 2000*: 515-518. Marseille.

Bendi, C. 2000. Ceramica comune depurata. In J. Ortalli, P. Poli and T. Trocchi (eds) Antiche genti della pianura. Tra Reno e Lavino: ricerche archeologiche a Calderara di Reno. *Quaderni di Archeologia dell'Emilia Romagna* 4: 51-54.

Bienert, B. 2001. Katalog der Sammlung Jacob Plein-Wagner: Die römische Keramik. In M. J. Plein (ed.), *Eifelkeramik. Sammlung Jacob Plein-Wagner. Führer durch das Töpfereimuseum der Familie Plein in Speicher*: 17-60. Speicher.

Bouet, A. 2009. *Les latrines dans les provinces gauloises, germaniques et alpines.* (Gallia Supplement 59) Paris.

Brulet, R., Dewert, J.-P. and Vilvorder, F. (eds) 2001. *Liberchies IV. Vicus gallo-romain. Travail de Rivière. Fouilles du musée de Nivelles 1986/87 et 1991/97).* Louvain-la-Neuve.

Grassnick, M. 1992. *Bäder und hygienische Einrichtungen als Zeugnisse früher Kulturen.* München.

Hayes, J. W. 1983. The Villa Dionysos Excavations, Knossos: The pottery. *Annual of the British School at Athens* 78: 97-169.

Herbin, P., Loridant, F. and Ménard, R. 2004. Un hypocauste de Bavay (Fache des Neuf Fontaines). *Revue du Nord* 86: 37-52.

Hilgers, W. 1969. *Lateinische Gefäßnamen. Bezeichnungen, Funktion und Form römischer Gefäße nach den antiken Schriftquellen* (Bonner Jahrbücher Beiheft 31)

Klauda, M. 1986. Geschichte und Geschichten vom Nachttopf. Erstes Nachttopf-Museum der Welt.

Katalog Zentrum für Außergewöhnliche Museen. München.

Loeschcke, S. 1922. Tonindustrie von Speicher und Umgebung. *Trierische Heimatblätter* 1: 5-13.

Petznek, B. and Radbauer, S. 2010. Römische Nachttöpfe aus der Zivilstadt von *Carnuntum*. Ein Fundensemble von der sogenannten Weststraße. Mit einem Beitrag von R. Sauer zu mineralogischen und petrographischen Analysen. *Carnuntum-Jahrbuch 2008*: 51-91.

Richter, G. M. A. 1966. *The furniture of the Greeks, Etruscans and Romans*. London.

Smith, A. H. 1904. *British Museum. A catalogue of sculpture in the Department of Greek and Roman Antiquities* 3. London.

The meaning of *stercus* in Roman military papyri – dung or human faeces? Or: who is supposed to clean *this* shit up?

Kai Juntunen

From the world of Antiquity, we have relatively few literary references to the activities related to toilets, either their use, cleaning or emptying. Instead, we have plenty of references to the actual excrements, but these instances favour circumstances that are related to dung, while human faeces tends to be a rather rare subject. It is due to this scarcity of sources that so much emphasis is placed upon on the few instances we have that illuminate this aspect of ancient life. One such a case in point is the fragmentary papyrus presently held at the Bibliothèque de Genève (cf. Appendix I) containing partial duty rosters of a legion, probably *legio III Cyrenaica*, datable to the late Flavian period. One of the duties listed is that of one M. Longinus, whose daily obligations are referred to as something that is defined by the Latin term *stercus*.

The precise wording of the assignment is uncertain, and several reconstructions have been suggested over the years. The original editors Nicole and Morel (1900: no. 1) interpreted the line(s) to read 'in / stercuss', which was later modified by Lesquier (1918: 232) into 'ad / stercus' and eventually supplemented by Marichal (in *ChLA* I [1954]: 17–18) as 'ad / stercus 7(centuriae)'. Since then, Marichal's interpretation has been the one most scholars (Daris 1964: 58; Watson 1969: 225) have adopted, but some degree of uncertainty still remains regarding the proper reading, and so Fink (1971: 114n.32) has even suggested that the line(s) could possibly be read as 'vic(o) al(exandrino?) / stercus ..' with two strokes after the last word being undecipherable. What all scholars have been able to agree on is the term *stercus* in the text, which is clearly visibly in the otherwise very worn-out and fragmentary papyrus.

Marichal originally (in *ChLA* I [1954]: 18) suggested that the order should be translated as 'duty of cleaning of the Century', but as of present, the most common interpretation of this work assignment perceives it to mean toilet fatigues. The first one to make this assumption seems to have been Watson (1969: 73, 229), who took the work assignment to mean 'cleaning of latrines'. He did not provide any arguments on which he based his assumption, and thus one can only assume that it was based on the most common meaning of the term *stercus* which is excrement. Davies (1974: 316 and table B) has suggested the same interpretation for the work order, and he did not provide any arguments for his assumption either, or reference to Watson for that matter, thus suggesting that he came to the same conclusion independently. In a similar spirit, without giving any explanations for his reasoning, Fink (1971: 114n.32) has stated that the work order could mean either toilet fatigues or alternatively cleaning of the stables.

The accuracy of these interpretations can be questioned as they seem to be based purely on the assumption that here the term *stercus* refers to human excrement. Neither is it a straightforward conclusion that a term that has a denotation as excrement should be understood to refer to latrines in general. To properly understand the meaning of the duty assignment of M. Longinus we need to explore the precise semantic value of the term *stercus*, the structure and function of Roman military toilets, and the relation of this duty to the other types of work assignment in the roster(s). In this examination, the first issue that needs to be addressed is how the Romans perceived the term *stercus*, not only in the official Latin literature, but also in everyday use, and also which terms the Romans used when they spoke about human excrements.

Stercus – terminological analysis

In the Classical Latin literature, the term *stercus* is most commonly found in the works which deal with agriculture or veterinary medicine. Not surprisingly, in these works *stercus* is used to mean either dung or manure processed from dung (for the literature references, cf. Appendix II). On occasion, the term can also be found to mean fertilizer that was composed of composted vegetation only and thus being free of animal excrement. This over-representation of the term *stercus* in the works that deal with agriculture and medicine is partly due to the nature of the topic of these works – each dealing in some fashion with excrement or excrement based fertilizer – but also in part to the fact that many of these authors used and quoted their predecessors, thus causing multiple instances of similar sentences to appear in the extant Latin literature. There are only a few instances in the agricultural manuals when *stercus* was used to mean human faeces, but on these occasions the meaning of the term was always defined by adding the adjective indicating human origin (i.e. *stercus humani*). In medical manuals on the other hand, the term could be used independently to mean human faeces, but here the subject at hand was either human anatomy or the symptoms of illnesses in human body, which made the semantic value of the term explicit.

The term does not appear so often in other forms of literature, but when it does it seems to follow similar patterns. In works of philosophical, historiographical and poetic nature, the most common meaning for the term is dung, while in a few occasions it is also used in reference to manure. It can also be found to indicate different forms of refuse or filth, like industrial manufacturing waste products, general sewage waste, or accumulated dirt. As a general expression for filth, it can also found in Cicero, who quotes a slander of C. Servilius Glaucia, who is called the 'Filth of the Senate' (*stercus Curiae*). Only a few times the term can be found in such literature to indicate human faeces and on these occasions its meaning as such is defined again by the use of the adjective indicating human origin (i.e. *stercus humani*). Also, when speaking of dung or manure, it is obvious from the literary context that the authors often meant large deposited quantities by this term, although occasionally a more specific term *stercilinum/sterculinum* can also be found – but seemingly almost exclusively in the agricultural manuals – when the author is speaking of dung-pits or dung-heaps meant for the production of manure.

The use of the term *stercus* in literature can only illustrate how the educated class understood the term, while the vernacular meaning among the common people could have been slightly different. In this, both Roman onomastics and Latin epigraphy (for a more detailed analysis of the Latin epigraphic instances, cf. Appendix III) can help us to understand how ordinary people used it and whether there existed a terminological difference between human and animal excrements. In epigraphy, the term does not appear so often, but when it does, its use is quite uniform as it can be found essentially only in town regulations, private warnings, curses, or shop signs. In *Luceria* (mod. Lucera), the term appears in a regulation (*CIL* IX 782) that prohibits discarding of *stercus*, abandoning of corpses or performing sacrifices in honour of the dead within a specific space. Here, the meaning of the term seems to be general refuse or waste, as the regulation prohibits its outpouring or discarding (*fundo*), thus indicating that large quantities were being transferred from somewhere else (Bodel 1994: 31). Similar restriction can be found in a decree (*AE* 1985: 358) from *Cingulum* (mod. Cingoli), prohibiting the discarding (*fundo*) of *stercus* within a specific area. Again, the verb used indicates transportation of the waste from another locality and not a personal act of defecation.

A third regulation of the kind (*CIL* IV 10488) comes from Herculaneum and prohibits the throwing (*abicio*) of *stercus* in the area of the water tower on *decumanus maximus*. In his analysis of this regulation Schubring (1962: 241–244) concluded that it was meant against polluting the area of the water supply with any kind of waste, including through the acts of defecation (*caco*) and urination (*mingo*). Hobson (2009: 144) has understood it to mean solely the act of defecation, while Bodel (1994: 32) interpreted the regulation as set up against dumping *stercus* <u>into</u> the public water distribution tank (*castellum*). Such direct pollution of the water tank hardly seems possibly as the tanks were set well above ground on top of a colonnade. Neither is it easy to understand how the verb *abicio* – that has the general meaning of throwing, dropping and abandoning – could be associated with the act of personal defecation. It is more likely that the regulation should be understood to have been intended against the pollution of the area of the water reservoir tank and the adjacent fountain through the dumping any kind of waste or filth, such as the dumping of industrial waste or emptying domestic waste buckets on the streets (for this practice in the Roman world, cf. Liebeschuetz 2000: 54, 59–60; Scobie 1986: 421; Taylor 2015: 75). The intention seems to be preventing bacterial and viral agents from spreading diseases into the water supply, and also preventing any kind of organic matter from attracting rodents that would have quickly infested the water tank.

All these regulations clearly refer to general waste disposal within the city limits or sacred precincts. Here, the meaning of the term *stercus* must be taken to mean any form of waste, whether it was the dung on the streets left by beasts of burden, construction or workshop refuse, or domestic waste (including human faeces) originating from household cesspits. In a similar spirit, several official inscriptions and private graffiti (*CIL* IV 7038, *CIL* VI 31614, *CIL* VI 31615, *CIL* VI 40885) were set to remind those who worked in the waste management business that the city regulations demanded the transportation of *stercus* (whether dung or other refuse) outside of the city limits. Two inscriptions (*CIL* VI 31577, *ILS* 8207b) make a further notification that one should not pile *stercus* on the graveyards (just outside the city gates) either. These regulations and notifications testify how common it must have been for private individuals to dump domestic waste in the relative vicinity of one's residence and for the professional sanitation workers to deposit larger quantities of waste temporarily in secluded areas within the city limits, instead of transferring it all the way outside the city limits into the properly defined areas.

Roman law provided clear rules of conduct for these professional sanitation workers and dung drivers or *stercorarii* (i.e. filth removers) as they were known collectively in the Latin language. The *lex Iulia municipalis* (i.e. *CIL* I² 593.66–67) which defined proper hours of the transportation within the city limits, also contained an exclusion for the transportation of dung and other waste (Hobson 2009: 99; Johnson *et al.* 1961: 94–95 [no. 113]; Liebeschuetz 2000: 53; Scobie 1986: 408), while other local decrees could define locations within towns were transportation was restricted. An inscription (*CIL* XII 2462a-b) from *Aquae* (mod. Aix-les-Bains) contains such a prohibition, against crossing the local *campus pecuarius* with a vehicle, and among the defined instances of movement, the transportation of *stercus* into a *sterculinum* is also mentioned. The specific definition of a product and its end location can be only understood as transportation of dung into a dung-heap or general waste into a dumpsite. The business opportunities created by the accumulation of waste within urban areas is also demonstrated by graffiti from Herculaneum and Pompeii (*CIL* IV 656, *CIL* IV 1754, *CIL* IV 10606) put up by entrepreneurs advertising their services, whether it was the removal of *stercus* by emptying domestic cesspits, removal of other accumulated waste, or selling and transporting of *stercus* presumably to a field as manure (Scobie 1986: 414; Taylor 2015: 81–82). The meaning of the term *stercus* in these advertisements can only be understood as either general waste or manure.

Another class of inscriptions employing the term *stercus* are the warnings and curses against desecrators. These inscriptions provide quite explicit wording of the deeds that are prohibited and against which the anger of the gods is being called down upon. The wording used often states that one is not to either defecate (*caco*) or urinate (*mingo*) in the described location (*CIL* IV 2357, *CIL* IV 7716, *CIL* IV 8899, *CIL* VI 13740, *CIL* VI 29848b), but occasionally there is also a separate addition against the deposition of *stercus* in the area (*CIL* III 1966), which is clearly separated from the acts of human defecation and urination, thus making it clear that dung (or general waste) is meant by the term (cf. Varrone 2016: 122n.27). The only other occasion that the term *stercus* can be found in epigraphy comes from a later inscription from Bregovina (*IMS* IV 114) that seems to repeat a modified verse from the Psalms. Here the verse reads: 'Qui pauperem ae stercore [elevas] / d(omi)ne nos (h)umiles servos [adiuva(?)]' ('You, Lord, who [raise?] the pauper from the dust, [help?] us your lowly servants'); in which the semantic value of the term *stercus* is either soil, dirt or dust, instead of literal excrement; the purpose being to illuminate how God saves also those of the most humble status.

The term *stercus* also appears as a root in a number of Roman names (for details cf. Appendix IV). Many of these, such as Sterceius, Sterceia, Sterculus, and the late third and fourth century forms such as Stercorius seem to have been primarily used by the common people, such as servants and slaves, but a few cases can also be found among the upper classes. There is no indication that any of these names would have been associated with human excrements, as we have no cognomina derived from the other words meaning explicitly faeces, but instead they seem to belong to the category of uncomplimentary cognomina. Thus, the semantic value of these names seems to have been associated with the concept of 'filthy', and these names were most likely understood by the Romans in a same spirit as the more famous names such as Crassus ('fatty') or Brutus ('dull-witted').

It would seem that the term *stercus* was most commonly understood as a generic term for any kind of waste, filth or dirt. The Latin language on the other hand does contain terms that were used almost exclusively of human excrements. Such terms as *fimus* and *merda* (Adams 1982: 231–244) can be found both in literature and epigraphy; and in both fields their use is most commonly human faeces. In the few preserved instances where the term *stercus* is used to describe human excrement, it is clearly defined so with an appropriate adjective (*stercus humani*), or it is used in a clinical context, such as in medical manuals. Nevertheless, it should be pointed out that the term always means the actual waste (whether it was human faeces, manure, or filth) and never the place, or a vessel into which such things were deposited.

Waste in military camps and the daily functions of the troops

Just like the ancient towns, the Roman military encampments were filled with life and activity, which generated various forms of waste as a by-product. The numerous workshops, everyday food preparation and the different animals that the troops required all contributed to the sanitation of the forts and their surroundings. In permanent camps, horses seem to have been the only animals allowed to be kept within the forts. Package animals (oxen, mules and donkeys) belonging to the army were probably kept in enclosures adjunct to the forts, while the keeping of other animals (sheep, goats, pigs, chicken etc. for dietary needs) was most likely outsourced to the civil settlements (*canabae/vici*) that quickly grew outside of every fort and legionary fortress.

As M. Longinus [...] seems to have been serving in *legio III Cyrenaica*, a unit that was based at the legionary fortress of Nicopolis just outside of Alexandria at the time when the papyrus was written, the only excrements he should have needed to deal with inside the fortress originated either from horses or humans. The composition and compaction of these two types of excrements is quite different and would necessarily have been required to be dealt with in different manners. Horse dung is in large part undigested grass. As such, it is usually relatively dry and can be piled and stored next to human habitation for extended periods of time. In comparison to other animal excrements, or human faeces for that matter, horse dung does not contain many bacterial agents harmful to humans, nor does horse dung emit quite as unpleasant a smell as other animal excrements do. In fact, most stables throughout history would have had a (horse) dung pile just outside the actual building, and with proper storage methods (by mixing the dung with dry hay or sawdust) such piles could be kept in place for months.

Human faeces on the other hand tends to be very soft, produces a repellent odour and is a source for a number of bacterial and viral agents that are harmful to humans. For these reasons it has always been essential to remove human excrements from the confines of human habitation as fast as possible. In Roman military forts this seems to have been accomplished by multiple different ways, the solution depending of the soil type, the intended user of the facilities (officers or troops) and atmospheric conditions. Most of the excavated Roman forts are located in the Northern provinces, and as such, their designs undoubtedly reflect the physical conditions of the colder climate. The officers had their own private latrines, while the needs of the troops were taken care of by means of larger toilet facilities. Some of these seem to have been built against the barracks buildings or sometimes even inside the barracks (Davison 1989: 233–237), the known cases originating primarily (but not exclusively) from the building periods when wooden structures were in use. The known structures seem to have been very simple in design, usually not more than depressions in the ground where latrine tubs (*lasana*) were held, with a possibly wooden structure above, and assumedly emptied on daily basis. The officers' latrines in their own houses on the other hand were often connected to drains if the fort had them, and in some cases the troop latrines by the barracks also followed a similar design.

The evidence suggests that by the second century, when most forts had been rebuilt in stone, faeces were dealt with by building multi-seated toilets (*foricae*) or a series of single-seated toilets (*latrines*) next to or against the wall on the sloping side of the fort (Hobson 2009: 35; Johnson 1983: 211). It is uncertain whether this design co-existed for some time with the barrack latrines or whether the latrine buildings by the wall line substituted the older barrack latrines – assuming that the forts had even had them in the first place. The known examples of such toilets seem to have been built in a rather similar fashion as their urban counterparts, namely having a connecting trough collect the excrements and leading the waste into a sewage pipe that ran out of the fort. In the excavated cases, there tends to be a water tank connected to the toilets, or they are located at the convergence of the fort's drains (Dixon and Southern 1997: 95–97; Goldwater 2011: 136–137; Hobson 2009: 35–41; Johnson 1983: 209). Clearly, the toilets were flushed either by using the tanks or the water in the drains, most likely multiple times a day to ensure the fluidity of the waste and to reduce the odour. Outside of the forts, the sewage pipes often led into septic tanks (Johnson 1983: 213), and the need to periodically empty such cesspits would have depended on the absorptive quality of the surrounding soil. In the few cases when no water supply for the toilets was available, the wall-line latrines seem to have used removable tubs or buckets instead (Johnson 1983: 213; Richmond 1968: 87), the design being practically quite similar to the earlier barrack latrines.

Nicopolis, a legionary fortress founded in the Augustan era, was undoubtedly connected to the same water supply system as the neighbouring city of Alexandria and thus would not have had any lack of running water. Most likely, the toilet facilities were built along the wall line by the end of the Flavian era and provided with a similar water flushing system as was used in the Northern provinces. Also, taking in consideration the heat in Egypt, it is doubtful that the Roman soldiers would have tolerated the stench of barrack latrines for over a century, no matter how much better ancient people were accustomed to strong odours. As of present, the regulations for public buildings in the US and Europe tend to require a single toilet seat for every 50 or 100 people. In ancient times, the relative need would not have been much different, and so an auxiliary fort should have sufficed with a single 10-to-20 seater toilet, with additional toilets outside of the fort in the bath house and separate toilets for the sick in the hospital building. A legionary fortress would accordingly have required ten times that number of seats. In addition, each barrack undoubtedly would have had urinals or portable pots (*matellae*) for urinating and possibly a few chamber pots (*lasana*) for the occasional nightly need to defecate (Davison 1989: 235–236; Goldwater 2011: 138; Wilson 2011a: 95f.).

In his study about Roman sanitation Scobie (1986: 409) refers to cesspits with the Latin term *sterquilinia*, but there are no primary sources were this term would have been used to indicate enclosed cesspits. Instead, the term *sterquilinum* (in its various spelling forms) is used exclusively of open-air dung-heaps and middens. The statutes of Roman law that concern public sanitation are equally vague about the terminology of specific structural entities. The Digest (43.23.9) speaks of individual's right to connect private sewers (*cloaca privata*) into public sewage system (*cloaca publica*). The indication of the law seems to be that these private 'sewers' had sometimes existed as independent entities prior to their connection to the public sewage system, and thus the term *cloaca privata* seems to have meant the household construction that included domestic cesspits and the pipes that led into them. Then again, the Digest also states (*Dig.* 43.23.1.6) that with 'sewage' (*cloaca*) both a channel tube (*tubus*) and a pipe (*fistula*) is meant, but no specific term for cesspits is provided. The only source that seems to make a more direct reference to urban cesspits is Juvenal, who in his *Satires* (6.602–

603) makes a jest of infants rescued from filthy cisterns. The term he uses to describe an artificial urban cesspool or cistern is *lacus* (Brown 1994: 195; Watson and Watson 2014: 264), a term that among other things is often used to describe constructed underground enclosures (tanks, cisterns, basins), or pit (i.e. hell) to where the lost souls are cast (cf. Vulg. *Psal.* 29.4).

If we return to the roster that formed our point of departure, the appointments of other soldiers are given by the physical location in which they served (for details cf. Fink 1971: 112–114). Thus, men were assigned to guard the gates or the headquarters (*statio portae/ principis*), work in the armoury (*armamentarium*), the baths (*ballio/balnio*) or (either cleaning or patrolling) the side streets of the camp (*strigae*). Some men were on detached duty, either as an escort for a tribune or a centurion, working at the limestone quarry or kilns (*ad cuniculos calcarios/cis*), while others were given leave or they were present at their *centuria*. Thus, the work assignment '*ad stercus [..]*' should in a similar spirit be understood to mean a physical location. As it has been pointed out, the term *stercus* refers strictly to the actual waste, not to the place or vessel used to hold or store it, and thus if Longinus was assigned to work at the toilets, the duty should have been recorded as '*ad foricae/latrines*'.

The structure and functionality of toilets that were flushed with water do not encourage the association of M. Longinus' duties with latrines either. The precise flushing mechanism of these toilets is uncertain, but they could have been simple enough for every soldier to operate them when the situation so required. Thus, there would not have been any need for a latrine orderly to serve by the toilets the whole day as some of the present interpretations suggest, and in such a case the order should have been given as indicating the physical location of the duty (i.e. '*ad foricae/latrines*'). Neither does it look likely that the work order could have meant the emptying of chamber pots either, as the used terminology cannot mean the vessels in use in the barracks (i.e. *matellae* and *lasana*) and as such a task would hardly require a whole day's work. In a similar fashion, the association of the duty with the emptying of cesspits seems equally unfounded, as the proper Latin term for such constructions would seem to have been *lacus*. One should also keep in mind that there were slaves and other servants in the Roman army, who tend to make their appearance in history only during campaigns when they are mentioned of taking care of the package animals and other similar tasks (cf. Dio 36.9.3; Jos. *BJ* 3.68–69). In peace-time, these same servants undoubtedly took care of a number of minor menial tasks, such as emptying of waste buckets and clearing of cesspits. As the term *stercus* can only mean physical waste, or a pile of waste, we should look for a large accumulation of waste in or near the encampment.

As mentioned above, horse dung can be and usually was stored in the open for extended periods of time. It is most likely that there was a selected location by the stables to where the stable boys removed dung and used hay from the stalls on a daily basis. Also, as many of the work orders in the papyrus are given in an abbreviated forms which do not always follow the usual standards, there is a possibility that the term *stercus* was abbreviated in a similar fashion as the term *armamentarium* on another line (i.e. III.5: *armamenta*). This leaves an interesting option that the final letter of the term that has been read as *stercus* actually belongs to another word which would then begin with a letter S. The second letter of this following word would then seem to be T, which is rather similar in appearance to the sign used for *centuria* (7), thus suggesting that the work assignment of Longinus should actually be read as '*ad stercu(s/linum) st(abuli)*' or 'dung pile by the stables'. Longinus' duty could thus have been either to remove the collected dung pile from the camp, or to supervise the stable boys doing the actual heavy lifting. The term *stercus* could also refer to the larger waste dump outside of the fortress, to where refuse from workshops, dung from stables and other waste was transferred. Just as any larger waste dumps, such an accumulation would need to be levelled from time to time. Whether a dung pile or a general waste dump, both installations could easily be referred to by the term *stercus* in Latin, and thus both could easily be understood as the physical location of the duties of M. Longinus.

Conclusion

The terminological analysis of the semantical value of the term *stercus* would seem to strongly contradict the popular belief that M. Longinus work assignment had anything to do with toilets. The term seems to have explicit value as waste in general sense, but never meaning the container or the location where this waste was deposited. The structural design of toilets in Roman forts in the late Flavian era and the tendency of the military rosters to differentiate the work orders by the physical location where each assignment took place, would further seem to contradict this popular belief. Instead, the duties of M. Longinus seem more likely to have been associated with the general waste management of the Roman fortress, whether it was the removal of the accumulated dung from the stables or clearing the whole encampment's waste dump.

Appendix I – Papyrus *ChLA* I 7a,b

Various aspects of this fragmentary papyrus (*ChLA* I 7a,b) have been under a prolonged debate ever since its original publication. The papyrus contains elements which make it certain that it belonged to a legionary archives in the late Flavian era, but the earliest commentators were unsure whether the legion in question was *legio III Cyrenaica* or *legio XXII Deiotariana* – the two legions based at Nicopolis at the time. One of the fragmentary rosters on the *verso* side of the papyrus contains a partial name, which Fink (1971: 107, 210-211) has reconstructed as *legio III Cyrenaica*. Even the very name of the soldier M. Longinus, whose duties are presently under a scrutiny, has received various interpretations, his illegible cognomen been variously read either as A[...] (Nicole and Morel 1900: no. 1), Au[...] (Marichal in *ChLA* I 7a,b [1954: 17-18]) or Ap[...] (Fink 1971: 110, 114). For details and bibliography regarding the different sections in the papyrus, cf. *ChLA* I (1954), 12-18; Fink 1974: 106-119 (nos. 9-10), 167 (no. 37), 210-212 (no. 58), 243-249 (no. 68).

Appendix II – Stercus in Latin Literature

The following literary references will be limited primarily to those included in the *Bibliotheca Teubneriana Latina* online database, and which were written by the mid fourth century CE so that they can be seen to reflect the semantic values of the Latin in use at the time when the papyrus fragment presently under examination was written.

Dung

In agricultural and medical works:

Cato, *Agr.* 7.3, 31.1 (generic), 36 (pigeon), 102 (swine), 151.2 (goat or sheep), 161.4 (sheep).
Celsus, *Med.* 5.8 (generic), 5.5 (lizard), 5.12 (pigeon), 5.27.8 (goat), 6.18.5 (sheep), 5.18.15 (ape).
Columella, *Arb.* 10.5, 17.2 (generic), 23.1 (swine).
Columella, *Rust.* 2.14.4, 2.15.5, 3.15.5, 5.9.14, 6.3.1, 7.4.6, 7.6.6, 8.5.19, 8.9.4, 10.vers.81, 11.2.92, 11.3.46 (generic), 8.8.6 (dove), 2.9.9, 4.8.3, 11.2.87 (pigeon), 8.1.2 (fowl), 8.3.7, 8.3.8, 8.7.2 (hen), 5.9.14, 11.2.87 (goat), 5.10.15 (swine), 11.3.12 (ass), 6.27.12, 6.30.6, 6.30.8 (horse), 9.14.8 (moth).
Gargilius Martialis, *Curae boum* 13 (rabbit), 15 (rooster), 19.
Pliny, *Nat. Hist.* 17.196, 19.177 (generic), 29.143, 35.46 (pigeon).
Scribonius Largus, *Comp.* 127 (mountain deer).
Varro, *Rust.* 1.19.3, 1.23.3, 1.38.1, 1.38.2, 2.7.11, 3.6.5, 3.8.3.

In other forms of literature:

Cicero, *Div.* 1.57.
Horatius, *Epod.* 12.11 (crocodile).
Hyginus, *Fab.* 30.7 (ox)
Iuvenalis, *Sat.* 14.64 (dog).
Lucretius, *Nat.* 2.871.
Minucius Felix, *Oct.* 24.5.

Petronius, *Sat.* 43.
Phaedrus, *Fab.* 1.27.
Plautus, *Asin.* 424.
Porphyrio, *Comm. in Horat. Epod.* 12.8 (crocodile).
Seneca, *Quest. Nat.* 3.26.5.
Suetonius, *Vit.* 17.2 (quoted by Eutropius, *Brev.* 7.18.5).
Tertullian, *Ad Mart.* 4.5; *De Pud.* 9.21 (dung/filth).
Valerius Maximus, *Mem.* 1.7.ext.10.
Varro, *Ling.* 6.32.

Manure

In agricultural and medical works:

Apicius 7.16.1, 8.6.6.
Cato, *Agr.* 2.3, 5.8, 10.1, 29, 31.1, 33, 37.3, 39.1, 48.2, 65, 114.1, 161.1, 161.4.
Columella, *Arb.* 5.4, 6.2; *Rust.* 2.5.1, 2.5.2, 2.9.15, 2.10.6, 2.10.7, 2.10.23, 2.10.27, 2.12.9, 2.13.1, 2.13.3, 2.14.1, 2.14.6, 2.14.9, 2.15.1, 2.15.2, 2.15.4, 2.17.2, 2.17.7, 2.21.3, 3.11.4, 3.11.9, 4.32.5, 5.9.9, 11.2.18, 11.2.86, 11.3.13, 11.3.28, 11.3.53.
Pliny, *Nat. Hist.* 17.127, 18.120, 18.194, 18.227, 18.322, 19.131, 19.148.
Varro, *Rust.* 1.2.21, 1.19.3.

In other forms of literature:

Fronto, *Ep.* 3.3.
Macrobius, *Sat.* 1.7.25.
Tertullian, *Ad Nat.* 2.9.20.
Varro, *Ling.* 5.139.

Composted Vegetation:
In agricultural and medical works:
Cato, *Agr.* 37.2.
Columella, *Rust.* 2.10.1, 2.14.5.
Pliny, *Nat. Hist.* 17.55.

Human Faeces

In agricultural and medical works:

Columella, *Agr.* 21.2 (stercore humano).
Columella, *Rust.* 5.10.10, 5.10.15 (stercore humano).
Celsus, *Med.* 2.3.6, 2.12.2B, 4.20.1, 4.22.1, 5.26.16, 5.26.17, 7.18.5, 7.20.2.
Scribonius Largus, *Comp.* 118.

In other forms of literature:

HA, *Com.* 11.1 (humana stercora).
Tertullian, *Adv. Psych.* 6.1 (interioris hominis ... stercoribus).
Vitruvius, *De arch.* 10.16.10 (stercoris humani).

Generic Waste

Apuleius, *Flor.* 14 (burden of filth or manure)

Cicero, *Orat.* 3.164 (quoted by Quintilianus, *Inst.* 8.6.15). It is uncertain what kind of filth is meant, but as the slander seems to have been part of an attempt to oust Glaucia from the Senate, the indication could have been cleaning of domestic waste from a house. For closer analysis of this quote, cf. Morgan 1974: 317–319.
Columella, *Rust.* 12.18.3 (waste accumulated in the wine cellar), 12.51.1 (dirt on wine grapes).
Firmicus Maternus, *Math.* 8.20.1 (waste from latrines and sewages).
Lucilius 398–399 Marx (waste or dung in military camps), 1018 Marx (filth of the ground, in contrast to human faeces [fimus] and pig excrements [sucerda])
Macrobius, *Sat.* 3.16.17 (sewage waste).
Phaedrus, *Fab.* 4.19 (trash pile sniffed by dogs for food).
Phaedrus, *Fab.* 4.25 (filth eaten by a fly).
Plautus, *Truc.* 556 (possessions treated as domestic waste). Although most translations of this play have interpreted the term as dung, it is clear from the context that indoor filth is meant.
Scribonius Largus, *Comp.* 48 (lead waste), 188 (iron waste).
Seneca, *Apocol.* 7 (filth of the lawyers being equal to the sewers of Augeas).
Seneca, *Dial.* 3.17.3 (in stercore suo destituti 'abandoned in his own filth [i.e. sweat, urine, excrements]).
Tertullian, *Adv. Marc.* 3.10.1, 4.14.5, 5.20.6 (dirt).
Varro, *Ling.* 6.32 (refuse from the temple). According to Festus (s.v. *stercus*, L 466), there was a special gate in Rome called porta Stercoraria through which this refuse (*stercus*) was expelled to the Tiber annually on June 15th. Ovid (*Fast.* 6.713) reports the same tradition, but refers to the temple refuse with the term purgamina. The precise nature of this refuse is discussed in detail by Holland (1961: 319-321).
Vitruvius, *De arch.* 10.16.7 (waste mixed with mud and water), 7.9.1 (industrial waste).

Dung-heap (Stercilinum/Sterculinum/Sterquilinum)

In agricultural and medical works:

Cato, *Agr.* 2.3, 5.8, 39.1.
Columella, *Rust.* 1.6.11, 1.6.21, 2.14.8, 7.5.8, 9.5.1.
Varro, *Rust.* 1.13.4, 1.38.3, 3.9.14.

In other forms of literature:

Seneca, *Apocol.* 7 (cock on a dung-heap).

Appendix III – Stercus in Latin epigraphy

Town regulations:

CIL IX 782 (Luceria). For a detailed analysis, cf. Bodel 1994: 4f., Petito 2004: 217f.;
AE 1985: 358 (Cingulum), Cf. Bodel 1994: 30–31;
CIL IV 10488 (Herculaneum). Cf. Schubring 1962: 241–244.

Stercorarii

Advertisements and graffiti:

CIL IV 656 (Pompeii).
CIL IV 1754 (Pompeii).
CIL IV 10606 (Herculaneum).

Regulations and notifications:

CIL I² 593 (Heraclea Lucania).
CIL VI 31577 (Roma).
CIL VI 31614 (Roma).
CIL VI 31615 (Roma).
CIL VI 40885 (Roma).
CIL XII 2462a-b (Aquae), For analysis, cf. Mangas Manjarrés *et al.* 2013/14: 287–288.
ICUR IV 12019 (Rome).

Warnings:

CIL IV 7038 (Pompeii). Hobson (2009: 144), Varrone (2016: 123) and Wilson (2011b: 148) have translated the term stercorarius here as 'defecator, shitter', but in this they are undoubtedly mistaken. The proper Latin term (cacator) denoting a 'defecator' can be found on a number of graffiti in Pompeii (*CIL* IV 3782; 3832; 4586; 5438; 6641; 7714; 7715; 7716; 8899), and thus instead of referring to someone in a dire need to empty his bowels, this inscription seems more likely to be another warning to the professional sanitation workers (stercorarii) not to leave the collected waste (in a carriage) in front of one's property, but to proceed to the town walls with their loads. For graffiti referring to the act of *caco* ('defecation'), cf. Hobson 2009: 143–146; Jansen 2011: 170–172; Varrone 2016: 122–128.
ILS 8207b (Verona). Schubring (1962: 244) understood the verb *facio* in here as referring to the act of defecation as the term generally means 'to do', but its semantic value is more in line with the sense of building or erecting, and thus the inscription seems to warn against erecting a pile of waste in the area.

Curses and Prayers

CIL III 1966 (Salona).
IMS IV 114 (Bregovina). The phrasing of the inscription 'Qui pauperem ae stercore [elevas] / d(omi)ne nos (h)umiles servos [adiuva(?)]' is comparable with Psalm 112.7: 'Suscitans a terra inopem et de stercore erigens pauperem' (Vulg. ex hebr.: 'suscitans de terra inopem et de stercore elevat pauperem'), Tertullian (*Adv. Marc.* 4.5): 'qui suscitat mendicum de terra et de stercore exaltat pauperem', and the Old Testament (*I Sam.* 2.8):

'Suscitat de pulvere egenum, et de stercore elevat pauperem'.

Appendix IV – Stercus in Roman Onomastics

The root of the term *stercus* can be found in a number of Latin names (cf. Kajanto 1965: 246). Some of these, such as Sterculus (*CIL* VIII 3404), and its feminine counterpart Stercula/Sterculia (*CIL* VIII 19640) are identical with the name of the Roman god of fertilisers (i.e. Sterculus, Sterculius, Stercutus, Stercutius, Sterculinius), and as such they were most likely semantically associated with manure. The precise semantic values of the other name forms, such as Sterceius, Sterceia, Sterceianus, Stercila, Stercorina (*CIL* III 9748), Stercorosa, Stercusia (*CIL* II 390; *CIL* II 392; *CIL* II²/5: 878) and Philo*stercus* (*CIL* VI 22804) are uncertain, but it does not look like any of these were associated with human excrements as we have no names derived from the other Latin terms, such as *fimus* or *merda*, that were used almost exclusively of human faeces. The convention of these name is also obvious, as they can be found from almost every corner of the Roman Empire and they seem to have been bestowed upon both the free born (or freedmen) and slaves alike (as did their Greek counterparts, cf. Pomeroy 1986: 158). In the later time periods, the name forms of Stercurius (*CIL* III 1871; *CIL* III 8549; *CIL* V 914; *CIL* V 1666; *CIL* VI 31845; *CIL* VI 26849) and Stercoria (*AE* 1986: 46; *CIL* III 2117; *CIL* III 2684; *CIL* III 9837; *CIL* VI 5167; *CIL* VI 13107; *CIL* VI 26850; *CIL* VIII 10613) became more common, and by the third and fourth centuries, such names could be found even among the upper classes (*CIL* VI 31845; *CIL* III 7494).

There has been a certain amount of debate regarding the origin and meaning of such copronyms. Some have argued that they were given to individuals who were abandoned as infants on dunghills and thus, these persons received their names from the place from where they were saved, but there is no clear connection between such names and one's origin, and in fact infants known to have been rescued from dunghills **often** carry theophoric or complimentary names (Kajanto 1962: 48-49; Pomeroy 1986: 157-159). Also, such names seem to have run in a family (at least in Egypt) and no disdain can thus have been associated with them (Pomeroy 1986: 158). In a similar fashion, the noticeable popularity of the later name forms in the Christian era has raised suggestions that the adoption of such derogative names reflects Christian sense of humility or the abuse Christian people received from their pagan neighbours, but there is no clear evidence that such names would have been more popular among the Christian population than the pagan one (cf. Kajanto 1962: 45f.). Also, Roman names do not tend to be pejorative, but they can often be uncomplimentary. There are plenty of famous examples of Roman sense of humour represented in the onomastics, such as Crassus ('fatty') or Brutus ('dull-witted'), and so the names derived from the term *stercus* should more likely be associated with a more generic semantic value of the term such as 'filthy', and consequently not have anything to do with either human faeces or excrements in general.

Bibliography

Adams, J. N. 1982. *The Latin Sexual Vocabulary*. London.

Bodel, J. 1986. Graveyards and Groves. A Study of the Lex Lucerina. *AJAH* 11 1986 [1994]): 1-133.

Brown, R. D. 1994. 'The Bed-Wetters in Lucretius 4.1026. *HSCP* 96: 191-196.

Daris, S. 1964. *Documenti per la storia dell'esercito romano in Egitto*. Milano.

Davies, R. W. 1974. The Daily Life of the Roman Soldier under the Principate. *ANRW* 2.1: 299-338.

Davison, D. 1989. *The Barracks of the Roman Army from the 1st to 3rd Centuries A.D.* Oxford.

Dixon, K. R., Southern, P. 1997. *The Roman Cavalry. From the First to the Third Century AD*. Abingdon.

Fink, R. O. 1971. *Roman Military Records on Papyrus*. Cleveland.

Goldwater, A. 2000. Users of the Toilets: Social Differences: (9.3.) The Roman Military in Britain. In Jansen G. C. M., Koloski-Ostrow, A. O., Moormann, E. M. (eds), *Roman Toilets: Their Archaeology and Cultural History*: 135-139.

Hobson, B. 2009. *Latrinae et Foricae. Toilets in the Roman World*. London.

Holland, L. A. 1961. *Janus and the Bridge*. Roma.

Jansen, G. C. M. 2011. Cultural Attitudes: Interpreting Images and Epigraphic Testimony. In Jansen G. C. M., Koloski-Ostrow, A. O., Moormann, E. M. (eds), *Roman Toilets: Their Archaeology and Cultural History*: 165-176. Leuven – Paris – Walpole.

Johnson, A. 1983. *Roman Forts of the 1st and 2nd Centuries AD in Britain and the German Provinces*. London.

Johnson, A. C., Coleman-Norton, P. R., Bourne, F. C. 1961. *The Corpus of Roman Law. Vol. II. Ancient Roman Statutes*. Austin.

Kajanto, I, 1962. On the Problem 'Names of Humility' in Early Christian Epigraphy. *Arctos* ns. 3: 45-53.

Kajanto, I. 1965. *The Latin Cognomina*, Helsinki.

Lesquier, J. 1918. *L'armée romaine d'Égypte d'Auguste à Dioclétien*. Cairo.

Liebeschuetz, W. 2000. Rubbish disposal in Greek and Roman cities. In Raventós, X. D., Remolà, J.-A. (eds), *Sordes Urbis. La eliminación de residuos en la ciudad romana*: 51-61. Roma.

Mangas Manjarrés, J., Álvarez Rodríguez, A., Benítez Ortega, R. 2013/2014. 'Casa / Casae' en el Occidente romano. *HAnt* 37/38: 271-298.

Morgan, M. G. 1974. Glaucia and Metellus: a note on Cicero, De Oratore II 263 and III 164. *Athenaeum* 52: 314-319.

Nicole, J., Morel, C. 1900. *Archives militaires du Ier siècle.* Genève.

Petito, A. 2004. Vicissitudini di un'epigrafe (CIL IX, 782 = I², 401). *ZPE* 147: 217–224.

Pomeroy, S. B. 1986. Corponyms and the Exposure of Infants in Egypt. In Bagnall, R. S., Harris, W. H. (eds), *Studies in Roman Law in Memory of A. Arthur Schiller*: 147–162. Leiden.

Richmond, I. 1968. *Hod Hill. Vol. 2. Excavations carried out between 1951 and 1956 for the Trustees of the British Museum.* London.

Schubring, K. 1962. Epigraphisches aus Kampanischen Städten. *Hermes* 90: 239–244.

Scobie, A. 1986. Slums, Sanitation, and Mortality in the Roman World. *Klio* 68: 399–433.

Taylor, C. 2015. A Tale of Two Cities: The Efficacy of Ancient and Medieval Sanitation Methods. In Mitchell, P. D. (ed.), *Sanitation, Latrines and Intestinal Parasites in Past Populations*: 69–97. Farnham – Burlington,

Varrone, A. 2016. Newly Discovered and Corrected Readings of iscrizioni 'privatissime' from the Vesuvian Region. In Benefiel, R., Keegan, P., *Inscriptions in the Private Sphere in the Greco-Roman World:* 113–130. Leiden.

Watson, G. R. 1969. *The Roman Soldier.* Bristol.

Watson, L., Watson, P. 2014. *Juvenal: Satire 6.* Cambridge.

Wilson, A. 2011. Urination and Defecation Roman-Style: Chamberpots. In Jansen G. C. M., Koloski-Ostrow, A. O., Moormann, E. M. (eds), *Roman Toilets: Their Archaeology and Cultural History:* 95–97. Leuven – Paris – Walpole.

Wilson, A. 2011. The Economy of Ordure: The Uses and Value of Excrement. In Jansen G. C. M., Koloski-Ostrow, A. O., Moormann, E. M. (eds), *Roman Toilets: Their Archaeology and Cultural History*: 147–148. Leuven – Paris – Walpole.